A COST EFFECTIVE USE OF COMPUTER AIDED TECHNOLOGIES AND INTEGRATION METHODS IN SMALL AND MEDIUM SIZED COMPANIES

IFAC Workshop, Vienna, Austria
7 - 8 September 1992

Edited by

P. KOPACEK
University of Technology, Vienna, Austria

UStrath

Published for the

INTERNATIONAL FEDERATION OF AUTOMATIC CONTROL

by

PERGAMON PRESS

OXFORD · NEW YORK · SEOUL · TOKYO

UK	Pergamon Press Ltd, Headington Hill Hall, Oxford OX3 0BW, England
USA	Pergamon Press, Inc., 660 White Plains Road, Tarrytown, New York 10591-5153, USA
KOREA	Pergamon Press Korea, KPO Box 315, Seoul 110-603, Korea
JAPAN	Pergamon Press Japan, Tsunashima Building Annex, 3-20-12 Yushima, Bunkyo-ku, Tokyo 113, Japan

First edition 1993

Library of Congress Cataloging in Publication Data

A cost effective use of computer aided technologies and integration methods in small and medium sized companies: IFAC workshop, Vienna, Austria, 7-8 September 1992/edited by P. Kopacek. — 1st ed. p. cm.
"IFAC Workshop on A Cost Effective Use of Computer Aided Technologies Integration in Small and Medium Sized Companies"—P.
1. Computer integrated manufacturing systems—Congresses. 2. Automation—Congresses. I. Kopacek, Peter. II. International Federation of Automatic Control. III. IFAC Workshop on A Cost Effective Use of Computer Aided Technologies and Integration in Small and Medium Sized Companies (1992: Vienna, Austria)
TS155.6.C684 1993   670'.285   93-5238

British Library Cataloguing in Publication Data

A catalogue record for this book is available from the British Library

ISBN 0-08-042061-3

Printed in Great Britain

IFAC WORKSHOP ON A COST EFFECTIVE USE OF COMPUTER AIDED TECHNOLOGIES AND INTEGRATION METHODS IN SMALL AND MEDIUM SIZED COMPANIES

Sponsored by
International Federation of Automatic Control (IFAC)
- Technical Committee on Manufacturing Technology (MANTECH)

Co-sponsored by
International Federation of Information Processing (IFIP)
International Federation of Automatic Control (IFAC)
  Technical Committees on
- Computers (COMPUT)
- Economic and Management Systems (EMSCOM)
- Education (EDCOM)

Organized by
Austrian Centre for Productivity and Efficiency
Institute of Robotics, University of Technology, Vienna

Supported by
University of Technology, Vienna, Austria

International Programme Committee

P. Kopacek (A)	(Chairman)	F. Nicolo (I)
L. Basanez (E)		G. Olsson (S)
C. Bonivento (I)		J. Ranta (SF)
Th. Borangiu (RO)		J. Scrimgeour (CDN)
K. Iwata (J)		J. Somlo (H)
L. Javorcik (CSFR)		Jiang Xinsong (PRC)
N. Kheir (USA)		M. Zaremba (CDN)
L. Nemes (AUS)		

National Organizing Committee
M. Zauner   (Chairman)
J. Hähnel
H. Strasser
P. Kopacek

PREFACE

The first IFAC-Workshop on "Cost Effective Use of Computer Aided Technologies and Integration Methods in Small- and Medium sized Companies" has revealed the latest state of the art in this subject - very important for smaller countries. CIM is an efficient tool, especially for small- and medium sized companies to increase the flexibility in production. Various scientific events all over the world deals mainly with questions of CIM- soft- and hardware but mainly for larger companies. The demands of small- and medium sized companies in the field of CIM are quite different in many cases.

The objective of this workshop was to bring together end-users, manufacturers and (computer) control specialists to discuss possibilities in the field of factory automation. Solutions for product, process-design, production-design and control were presented. Technical criterias were discussed and economic justification methods were evaluated. According to the title the main issue was to present intelligent modular low cost approaches as solutions appropiate for small- and medium sized companies.

Two plenary papers as well as 26 regular papers are included in this preprints volume. These papers are from different fields related to control problems in CIM as well as other related topics for example computer aided manufacturing, computer aided quality insurance, computer aided planning, robotics, networks, software design and CIM-management. These papers give an overview about the actual state of the art of this field.

Finally it is a pleasure for me to thank all members of the IPC and NOC for the hard work in support of CIM'92. I would also like to thank the Austrian NMO, the Austrian centre for productivity and efficiency as well as the University of Technology of Vienna for the generous affords in the preperation of this symposium.

Vienna, October 1992

P. Kopacek

IPC Chairman

CONTENTS

CIM for small and medium sized companies

P. Kopacek and M. Zauner

University of Technology Vienna
Vienna, Austria

Scientific Academy of Lower Austria
Krems, Austria

Abstract: Since some years CIM is a headline in research of factory automation. The literature in this field was growing up dramatically and concepts as well as software packages for CIM components like CAD, CAP, PPS, CAM, CAQ/CAT were developed but only partially realized in companies. These commercially available packages are mostly suitable for large companies in distinct fields but usually not for small and medium sized companies in different fields. The Austrian industry is dominated by such companies with up to 500 employees.

Another problem for such companies mentioned before are the investment costs for computer soft- and hardware for the installation of a CIM system. These facts lead to so-called "island" solutions for some CIM components.

Therefore a "low cost" modular CIM concept especially for small and medium sized companies was developed. As a pilot project it was tested first in an Austrian medium sized company producing welding transformers. This concept is based on the hardware of a network of PCs (operating system MS-DOS) and a database computer (operating system UNIX). The realization of this solution is accomplished by using the C-programming language, Informix database and Informix-SQL as database query language, in order to create the custom applications with database management. With these tools, the structure of a relational database and the ANSI-standard Structured Query Language are provided. This concept is being described and an application example is discussed with special emphasis on problems of information techniques.

Keywords: Automation, CAD/CAM, industrial production systems, manufacturing processes, low cost CIM concept, data communication.

1. Introduction

CIM should be a tool for a totally computer aided production. Software packages for the so-called CIM components like CAD, CAP, CAM, CAQ/CAT and PPS are commercially available from various companies. Usually these packages are only partially suitable for small and medium sized enterprises. They offer a lot of features not necessary for the demands of small and medium sized companies, on the other hand some special features - absolutely necessary - are not included. These features are difficult to add in such packages. Furthermore software packages for CIM-components from various sellers require different hardware facilities as well as different operating systems. From the investment side small and medium sized companies have the problem to spend all the whole money for a CIM installation at the same time.

2. CIM COMPONENTS

According to common definitions, CIM is closely related with the components of:

CAD: includes all data processing activities, related to developement and construction, together with calculation and simulation (CAE)

CAP: planning of work, programming of NC - machines (results are basic data of the PPS - system).

CAM: includes the support of computers for the control of the working means during the production (machine tools, handling devices, transport and stock systems).

CAQ: describes computer supported planning and performance of quality control.

PPS: includes the use of computer aided systems for the organisatorial planning, controlling and monitoring of all processes.

For these reasons a modular "low cost" CIM concept was created especially with regard to the demands of small and medium sized companies. This development was supported by the Austrian Ministry of Science and Technology. From the side of hardware, the basic philosophy is to use PC´s (e.g. 386 compatibles) and a host computer for the database. The operating system for the PC´s us MS-DOS, for the host computer UNIX. All computers are connected by a LAN (ETHERNET).

3. THE MODULAR "LOW COST" CIM CONCEPT

The modular "low cost" CIM concept is shown in Fig.1. It uses two types of computers: A UNIX machine serves as a database and network server,

various MS-DOS computers (AT or 386) work as network stations with different tasks. For OCA special terminals are used.

Modules of the CIM system are:

Development:
- CAD workstations
- NC-programming workstations and simulation
- NC-programming workstations located at the machines

Production planning:
- PPS
- Production Control System (PCS)
- Group Control System (GCS)

Production:
- manual working places
- various NC-machines

other units:
- database (DB)
- Task Pool Manager (TPM)
- Operating Characteristics Aquisition (OCA)
- CAD/NC-fileserver

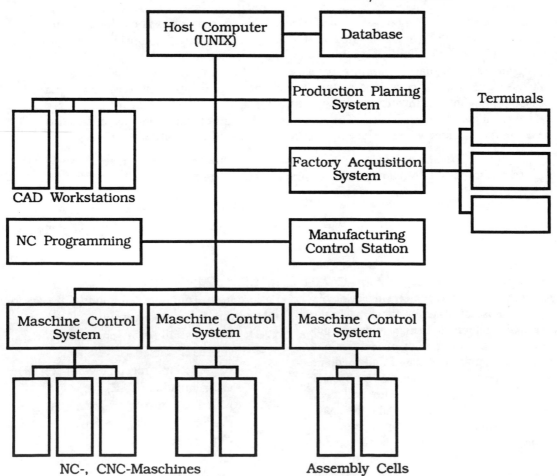

Fig. 1: "Low Cost" CIM Concept

Except the PPS-system, the personal oriented part of OCA-interpretation, and the CAD-systems all programs had to be developed.

The database-server and all workstations are connected by a local area network (ETHERNET). A second network (party-line) connects the OCA-terminals with the OCA-server, which works as an intelligent gateway between the two networks.

This concept has following advantages for small and medium sized companies:

- "low cost"
- possibility of stepwise realization
- easy adjustment of the software
- possibility to include some methods of AI

4. FLOW DESCRIPTION IN THE CIM-SYSTEM

Physical Connection

The database-server and all workstations are connected by a local area network (ETHERNET). A second network (party-line) connects the OCA-terminals with the OCA-server, which works as an intelligent gateway between the two networks.

Logical Connection

On the database and network server runs a Data Communications Managment (DCM) which allows different tasks to communicate. The database, processes and workstations have one well-defined interface to exchange data and messages. On top of DCM an SQL-shell enables

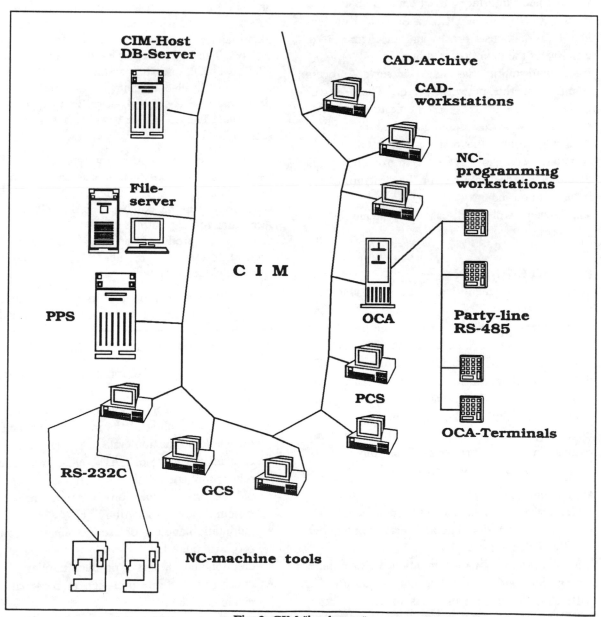

Fig. 2: CIM-"landscape"

high-level database processing. See Fig. 2 for an example of the data flow.

The diverse control systems recognize on their own, if the connection to the host server is possible and/or succeeded. They build up the connections self-acting and adjust the data between the different databases automatically.

With this solution the personell in the production is not stressed with additional data-processing activities, in order to keep the control system working.

5. CILIENT / SERVER CONCEPT

On the database and network server runs a Data Communications Managment (DCM) which allows different tasks to communicate. The database, processes and workstations have one well-defined interface to exchange data and messages. On top of DCM a SQL-shell enables high-level database processing. In Figure 3 an example of the data flow is shown.

Each workstation has its own server process running on the network server. These server processes handle the communication between workstations, processes and database.

As a pilot project this concept was installed in an Austrian company with approximately 500 employees, representative for the structure of the industry of our country.

The concept will be discussed by means of this pilot project.

6. START SITUATION

This company mentioned above produces 6 different basic types of fully transistorized welding transformers, which have been developed during the last five years. Due to some circumstances from these 6 basic types approximately 1400 different welding transformers have been derived and are now being produced. According to the demands of the market this number is still increasing today. This leads to a decreasing number in one distinct series. From this point of view there are serveral objectives of the company:

- decrease in time of pass
- improvement of capacity utilization factor
- to accomplish the huge and increasing number of varieties in products

According to the situation mentioned above there are special conditions, which are unique for this particular company. This leads to certain frame conditions, which need to be worked out. In our case, there were conditions like:

- machine tools, data processing systems and software packages in the factory exist in a wide variety of types and from different producers
- different standards and/or interpretations of existing interfaces
- interfaces are not compatible.

This leads to a solution, where the demands of the company require a CIM concept:

- which is future oriented (modern hardware: UNIX-, MS-DOS-computers, LAN's)
- which has new software techniques (relational database, client-server concept)
- which has a modular design.

Two years ago, at the start of the project, two CAD systems, one for the electronic construction and one for the electromechanical construction, a PPS-system and some machine tools were in use. Stand alone PCs were available for the CAD systems. The PPS system runs on a minicomputer (Philips P4000, operating system DINOS). On the production level 8 machine tools equipped with programmable controllers of different types were available. These "islands" were not connected together.

For the development of a new type of a welding transformer the construction was carried out on the CAD system. At the same time the basic data were stored in the database of the PPS-system. According to the printouts of the CAD-systems for some parts produced on computer controlled production machines, the necessary NC programs were developed manually.

If the PPS system scheduled a distinct machine tool for production these programs were loaded manually in the programmable controller.

The new concept should fulfil the following tasks:

- generation of the NC programs directly from the CAD system
- loading these programs into the machines
- optimal scheduling of the machines as well as the whole production cycle
- data management of CAD-, NC-, and various other production data
- bidirectional connection between the PPS-system and the production level
- automatic handling of product changes at the production level
- creation and support of product families
- generation and interpretation of personal and production data

- information about the current production state on different levels
- modular file service by use of optical disks
- modularity
- "low cost"

Starting with the product design data are generated in the CAD-systems. Engineers and designers create or change components or complete products. The CAD-files and descriptions of the components are stored on the CAD/NC-fileservers and in the database. Operational engineers at the PPS-system create working sheets for the components using these data. For this purpose they need additional information from protoyping at the production level. Workers and production engineers write and check NC-programs created from the CAD-data. The tested NC-programs and additional informations about them are stored on the CAD/NC-fileservers and in the database for further use by planning and production processes. In paralle the quality assurance division develops testing programs. When all these production tools are designed and tested a component is ready for production.

Production is initiated by an external order of complete products. The PPS-system creates lists of components and time scheduled for their production. Raw material reservations are done

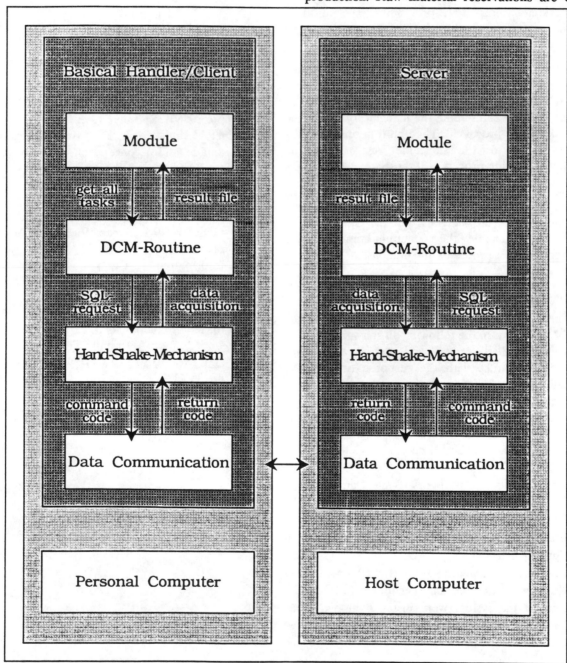

Fig. 3: Data Flow

and all necessary tasks are planned into the production cycle. The PPS-planning concentrates on the correct sequence of tasks, time estimations are not very expressive. The PPS-system interchanges production tasks with the TPM. The TPM handles short time planning on the production level and communicates with the PCS and GCS in the system. In addition all production oriented data from the OCA are interpreted. Therefore the TPM always mirrors the actual state of production.

The primary user interface for the TPM is the PCS. At the PCS a master plans tasks for fixed times and fixed GCS. The master can get additional information about the tasks, like working tools, NC-programs, special raw materials. The current state of each task can be displayed at the PCS.

The PCS sends tasks together with additional information and NC-programs to the selected GCS. The GCS stores these tasks. The worker at the GCS descides about the machine, on which the task will be processed. Then he transmits the NC-programs to the machine and starts the manufacturing process. All these states of the task are recorded by the OCA.

Each workstation has its own server process running on the network server. These server processes handle the communication between workstations, processes and database.

7. INSTALLATION AND TESTS

The installation started with the database and the OCA system. In the following all software modules were installed, tested and are now in use. Problems arose with the response time of the database. Only little adaption of the software was necessary, mostly due to additional demands of the operators. The new concept was accepted by the operators and they were able to work with it efficient in a very short time.

SUMMARY

Especially for the purposes of small and medium sized companies a modular, open "low cost" CIM concept was developed. As a pilot istallation it was tested in a typical medium sized company in Upper Austria. This project serves as a test case for a research emphasis supported by the Austrian Ministry for Science and Technology. Meanwhile parts of the system were installed in two other companies with high efficiency.

8. LITERATURE

Kopacek, P. and Fronius, K. 1989, "CIM Concept for the Production of Welding Transformers", Preprints Information Control Problems in Manufacturing Technology, (INCOM´89), Madrid, Vol. 2, pp.737-740.

Kopacek, P. 1989, "Automation of Small and Medium Sized Companies in Austria". Institute for Systems Engineering and Automation, University of Linz.

Hölzl, J. 1991, "Design and Developement of a CIM Concept for a Medium Sized Austrian Company", Institute for System Engineering and Automation, University of Linz, in German.

Kopacek, P., Frotschnig, A. and Kaçani V. 1991, "User Interface To A Cim-Database", Proceedings of the Second International Workshop on Computer Aided Systems Theory, (EUROCAST'91), Krems, "Lecture Notes in Computer Sciences", Vol. 585, pp.592-601, Springer 1992.

Zauner M. and Kopacek P. 1992, "Integration of data in a CIM-concept for small- and medium sized companies", 8th International Exhibition for Computer Aided Technologies with User Conference, Stuttgart, in German.

Kopacek, P., N. Girsule, J. Hölzl 1992, "A low cost modular CIM concept for small companies". Proceedings of the IFAC-Symposium on Information Control Problems in Manufacturing Technologies - INCOM'92, Toronto, pp. 188-192.

Frotschnig. A., P. Kopacek, M. Zauner 1992, "CIM for small companies". Preprints of the IFAC workshop on Automatic Control For Qualitiy And Productivity - ACQP'92, Istanbul, Vol. 1, pp. 35-41.

THE GRAI INTEGRATED METHOD: A TECHNICO-ECONOMICAL METHODOLOGY TO DESIGN MANUFACTURING SYSTEMS

B. VALLESPIR, C. MERLE and G. DOUMEINGTS

GRAI - Laboratory of Automatics and Productics (LAP), University BORDEAUX I, FRANCE

Abstract. Managing manufacturing taking in account objectives coming from commercial constraints such as flexibility, quality, etc.. makes todays industrial systems tremendously complex. Thus, designing such industrial systems leads to use structured methods in order to get finally a system working as expected at the beginning of the project. For this purpose, many methods are currently proposed. However, one of the main lacks of all these methods is that they propose no tool to evaluate during the project the technico-economical opportunities of the future system. Thus, the purpose of this paper is to present an integrated methodology to design CIM systems including a technico-economical tool supporting design choices in order to improve performances of the future manufacturing system.

Key Words. Design methods, CIM projects, evaluation, performance indicators, decision support systems.

1. INTRODUCTION

Today, according to the industrial competition, compagnies must manage their manufacturing systems according to various objectives such as productivity, flexibility, quality, reliability,...

The investment needed by the implementation of such new manufacturing systems is very important. It could be a disaster for the firms to make wrong specifications for these advanced manufacturing systems. Thus a CIM project must be supported by technical tools in order to enable to get good specifications according to the needs of the firm and economical tools to evaluate the rentability of investments. It is so necessary to use methods to perform this activity taking into account technical and economical points of view.

In this paper we will present a methodology for designing manufacturing systems: GIM (GRAI Integrated Methodology). Today this approach has the following main aspects:
• GIM is composed of reference models of a CIM system, modelling formalisms and a structured approach.
• GIM proposes several views of a CIM system: a functional view (to have a global understanding of the system to be built), a decisional view (to understand the system with an organization point of view), an informational view (to take choice about Information Technology options) and a physical view (to analyse the manufacturing options).
• These views are related to the skill of people involved in the project and are consistent.
• GIM owns an analysis phase enabling to take into account the initial status of the manufacturing system in order to well understand under which specific constraints the system operates and to avoid the re-designing of the parts of the system which are satisfying.
• The global process of design owns two main parts: the first one is user oriented, the second one is technical oriented in order to elaborate technical specifications.

In this framework we will present the improvement of GIM actually in process about the purpose of technico-economical evaluation. As a matter of fact we have already seen that current methods do not propose any tool for the technico-economical evaluation of the future system opportunities. To fill this gap, the current improvement of GIM are concerning a technico-economical tool supporting design choices during a CIM project. The main aspects of this tool are:
• The choices having to be taken all along the project are supported by a decomposition of objectives,
• Each design decision making is supported by a triplet <Objective, Design Variables,

Performance Indicator> which is the basis concept supporting the performance evaluation tool.

2. GIM: GRAI INTEGRATED METHODOLOGY[1]

Doumeingts *et al.* (1992), Vallespir *et al.* (1989, 1990).

The term "methodology" means here a consistent set of components (fig. 1) which are:
• a reference model globally and generically showing the structure of the system to be studied,
• one or some modelling formalisms enabling to build up the models of the system in order to study and evaluate it,
• a demarch leading step by step from an existing system to a future one taking into account evolution objectives and specific constraints,
• evaluation criteria with which the system can be evaluated in relation to various points of view (control consistence, reliability, etc.).

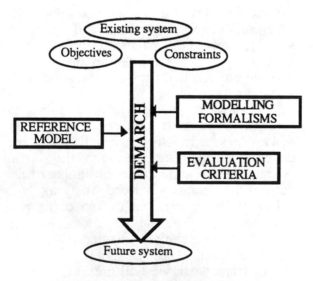

Fig. 1. Methodology components

2.1. Reference model

A reference model aims at modelling the invariable parts of the CIM system and the links between them. It is an a priori model which may be used as a basis for designing and implementing these systems. The GRAI model (fig. 2) aims at giving a generic description of what a manufacturing system is,

[1]: GIM was developped by GRAI in the frame of two ESPRIT projects: 418 (OCS) and 2338 (IMPACS).

focusing on the control part of this system. The sub-systems appearing in the GRAI model are:
• The physical sub-system (machines, workers, techniques, ...) transforming material flow.
• The decision sub-system split up into decision-making levels, according to several criteria, each one composed of one or several decision centres. The *operating system* dedicated to *real time* management can be set apart.
• The information sub-system is the link between decision and physical systems, with environment and aims also at transforming and memorizing information.

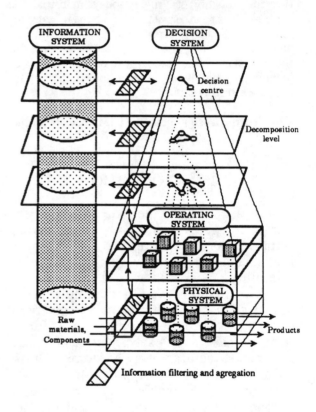

Fig. 2. The GRAI model

When focusing on control system, these considerations lead us to give prominence to a two dimensions decomposition:
• a fonctional one (vertical), defining the various functions of such a system,
• a hierarchical one (horizontal) in accordance to time criteria instituting decision-making levels characterized by the concept of horizon (time interval on which the result of the decision is relevant) and the concept of period (time interval after which the result of the decision must be questioned). These notions are a generalization of the three classical levels: long term, middle term and short term. This two dimensions decomposition enables

the concepts of decision centre to be defined: a decision centre contains as a whole the decisions made within one function and one hierarchical level.

2.2. The modelling framework

When designing a CIM system, because its high complexity, a first structuring is needed. Various concepts and models need so to be defined and to be built. In order to ensure the completeness, consistency and integration between the concepts and the models, we propose to define a modelling framework in which all the models needed for the analysis, design and implementation of CIM systems find their places.

The modelling framework has two dimensions: functional decomposition and abstraction levels.

Functional decomposition. According to the GRAI model, any manufacturing system may be splitted up into three systems: the physical system, the decisional system and the informational system. These three systems lead us to get three views. A view can be defined as a selective perception of a manufacturing system which concentrates on some particular aspect and disregards others. To these three views, we add a fourth one: the functional view. The functional view enables to get a model very simple to build showing the main functions of the manufacturing system and the flows (of any nature) moving between these functions. Another interest of these view is to define exactly the boundary of the study domain.

Abstraction levels. The modelling activity implies a simplification of a too complex reality. So, a model keeps only concepts and elements which will be necessary at the time of the model use. The introduction of the abstraction levels allows a 'stratified description' in the sense that our model is in fact constituted of several ones which integrate specific concepts. Practically, our model owns three abstraction levels.

Conceptual level. Made up without any organisational or technical consideration, it is the steadiest level and aims at asking the question 'What ?';

Structural level. It integrates an organisational point of view and aims at asking the questions 'Who ?', 'When ?' and 'Where ?';

Realizational level. It is the more specific level because it integrates the technical constraints of the studied case and enables the choice of real components.

Crossing these two dimensions gives the GIM modelling framework (fig. 3). On this figure we can see that the upper part of the modelling framework is user oriented while the other one is technical oriented.

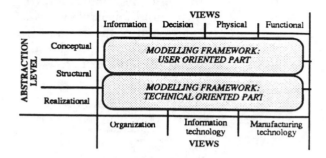

Fig. 3. The GIM modelling framework

2.3. Modelling formalisms

We will focus on the upper levels of the modelling framework (user orientation). The needs for each sub-domain are as follow:

Conceptual Information Model. The CIM is a description of all stable and 'natural' data of the organization, of their attributes and of links between them. The formalism used here is the Entity / relationship model.

Structural Information Model. The SIM describes the data structure in relation to the distribution of data and the computerized / manual choice. The formalism used here is also the Entity / relationship model.

Conceptual Decision Model. The CDM is a description of the decision making structure, links between decision levels, analysis of links between objectives, analysis of constraints, description of decision variables. The formalism used here is the GRAI grid at the global level and the GRAI nets at the detailed level.

Structural Decision Model. The SDM mainly enables the identification of decision makers, responsability and authority. It links decision makers and decision-making. The formalism used here is the GRAI grid at the global level and the GRAI nets at the detailed level.

Conceptual Physical Model. The CPM is a description of process and routes with physical flows between operations. The formalism is the actigram IDEFØ.

Structural Physical Model. The SPM gives information about time, work centres and operators, elements about linking and synchronization and indicates who does what. The formalism is the actigram IDEFØ.

2.4. GIM Structured approach

According to the user requirements of the future system, the goal of GIM is to provide specifications in terms of organization, information technology and manufacturing, which will allow to build this new system.
The structured approach of the method (fig. 4) has mainly four phases: initialization, analysis, design and implementation.

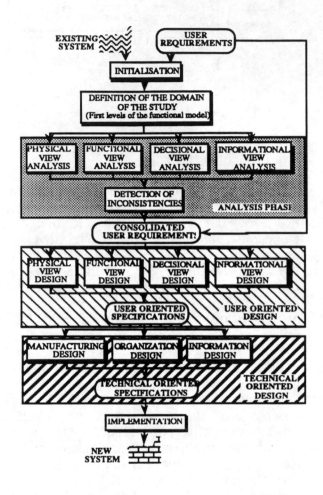

Fig. 4. GIM structured approach

We consider that the techniques used to build new manufacturing systems are currently very difficult to understand for the users of the future system in particular for the information technology domain. Besides, according to the amount of investment necessary to build a new manufacturing system, we need to be sure that the design of the new system reaches the objectives defined in the user requirement. The design of the new system must be validated by users before to start any developments or implementations.
Because of these remarks, we have split up the design phase into two sub-phases: the user oriented design which provides user oriented specifications understandable by users

and the technical oriented design which provides technical specifications necessary for the develoment and the implementation of the future system. Users validate the user oriented specifications to be sure that the design phase provides efficient solutions according to the requirements.

3. CONCEPTUAL MODEL FOR TECHNICO-ECONOMICAL EVALUATION

During a CIM project it is necessary to take choice according to various objectives which are sometimes in conflict. To take in account the complexity of the project the investment decision cannot be only based on classical methods such as pay-back, net actual value, rentability internal rate or neo-classical (costs advantages).
These approaches mainly focusing on short term rentability of the firm do not take into account neither the whole nor the modularity of CIM invesment (Merle, 1990). Besides these traditional approaches are designed for a stable environment and for production systems based on the standardization of production tasks. In front of moving environment and flexible manufacturing systems we must define new evaluation approaches.
For a problem of CIM investment the managers have in general way to choose among different solutions. Alternative solutions must be evaluated according to several criteria. To help this choice, we propose an evaluation model, both methodological and operational, which takes in account economical, financial, social, organisational, technical aspects and trading objectives at the various steps of the CIM project. This model is supported by the GIM structured approach.

3.1. Formalization of the model

We consider it is possible to give a hierarchy and to split up the overall objectives of the company into basic objectives. However, the manager must keep in mind the general objective. the priority of this objective must not be in conflict with the choice of basic objectives more precise and often defined for the short term. Such a hierarchical analysis procedure gives a simple and efficient way to help this decision-making. These objectives are measured by one or several performance indicators which can be quantify. Finally, these results are linked to design variables.
This process allows us to propose a structure

based on the triplet: <Objective, Performance Indicator, Design Variable>[2] (Merle, 1990, 1992) (fig. 5).

With this structure, each company can build by itself its own adapted evaluation model according to its requirements.

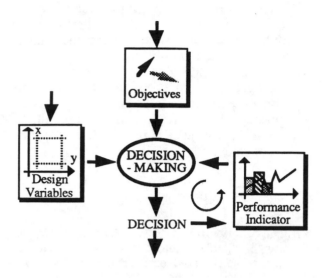

Fig. 5. The evaluation conceptual model

Figure 6 shows in detail the links between each components of the triplet. The CIM project manager gives a global objective (generally qualitative). This objective is decomposed in several levels (1, ..., i) of basic objectives (EOi1, ..., EOij). When the basic objectives are formulated and cannot be decomposed further, the definition of Performance Indicators (PIk1, ..., PIkm) for the k levels of results is possible. Each of them is calculated according to Design Variables (DV1, ..., DVn).

3.2. Definition of the main decisions to be made during a project, an example

It is obvious that the concept of performance indicator is strongly linked to decision-making. Thus, the definition of performance indicator is possible if decision centres are also defined within the project management structure.

Many of the main decisions to be made are concentrated in the passage from an abstraction level to the lower during the design phase.

To illustrate this kind of decision, let us

---

[2]: Another model based on the same triplet has already be used for the implementation of performance indicator for <u>operational</u> decision-making (ECOGRAI method) (Bitton, 1990, Vallespir *et al.*, 1990).

consider the passage from the conceptual model to the structural one during the design of a workshop (Vallespir *et al.*, 1991). For instance, three specific operations are needed to manufacture a product. At the conceptual level, we define these operations without representing how they are structured. At the structural level, we must define the structure that we will implement, taking into account various criteria and constraints. We can choose three individual machines or just one machine that can process the three operations. The first solution is more reactive in relation to failures, the second one is more flexible but less safe. Criteria taken in account here are reactivity and flexibility.

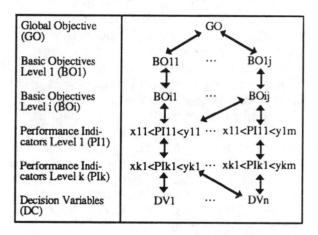

Fig. 6. Links between components of the model

In this example, the design variables are linked to the possible choice of machines available on the market; the performance indicators must be set up as follow: one reactivity indicator, one flexibility indicator. The final choice will be taken according to the objectives (not defined in the example).

3.3. Application to a subcontractor firm

We will analyze the CIM project of MECAMOUL to show the conceptual approach of the proposed evaluation model.

MECAMOUL is a subcontractor mechanical firm realizing a turnover of 30 MF, a net result after taxes of 2,1 MF and a cash flow of 3,6MF. It employs 50 persons.

The firm must resolve problems of lead time and quality. Subcontracting requires reactivity to answer mainly in terms of lead time. What is the position of the firm relating to the concurrents ? MECAMOUL aims to a medium size production. It also wants to develop the trade of manufacturing mould with high added value.

The basic criteria is to make mould machining represent 50% of the whole activity of the firm. Today, it is representing 20%. Its activity is directed towards the most important firms in France in the aeronautical and automobile industries and also with the constructors of equipment goods.

The requirements of this firm are average ordered pieces volume, type of subcontracted pieces, required quality level, order and delivery cycle and price level.

According to the firm analysis, we can present the articulation between objectives / performance indicators / design variables. The global objective of MECAMOUL is to improve the performance of the firm by increasing the activity of mould machining. Then, the basic objectives can be split up according to the differents sub-systems of the firm. For each basic objective, we define performance indicators. We can see some examples in the following matrix. The design variables will be: technical ressource type (machinery, tools), products type (semi-finished products, finish products), time type (work time, order cycle time), human ressources type (level of qualification, level of training), financial ressources type (investment cost, rate of rentability), economical ressources type (turn-over, net result, added value).

Basic objectives of level 1	Performance Indicators of level 1
Increase productivity	Work productivity rate Rate of work asset
Increase quality	Rate of scraps Rate of breakdowns Frequency of breakdowns
Develop an intensive commercial strategy	Research of new distribution networks Advertising expenditures of 150 000F
Answer to the technical specifications	Know-how Level of equipement qualification
Develop the part of high added value mould machining	Level of performance of men and equipment
Increase the market part	Increase of 10% on 3 years the mould machining Increase the number of customers
Elaborate a 2 years training plan	2000 hours for the design department

4. CONCLUSION

The use of methods for designing CIM systems is very helpful but are much more efficient if they propose a mean for the evaluation of the future system. Thus, within the framework of the GRAI Integrated Methodology (GIM), we propose to adopt a technico-economical approach based upon the relationship between objectives, performance indicators and design variables.

One of the main problem to solve now is the building of performance indicators. For this purpose, some research works are currently in process. In this framework, the results of the ESPRIT project 'FOF' (BRA-3143) may be very relevant.

5. BIBLIOGRAPHY

Bitton, M. (1990). ECOGRAI, Méthode de conception et d'implantation de systèmes de mesure de performances pour organisations industrielles. PhD Thesis, University Bordeaux I, 220 p..

Doumeingts, G., Vallespir, B., Zanettin M., Chen, D. (1992). GIM, GRAI Integrated Methodology - A methodology for designing CIM systems, Version 1.0. University Bordeaux I, LAP/GRAI, 62 p..

Merle, C. (1990). L'évaluation des projets productiques, contribution conceptuelle et méthodologique. PhD Thesis, University Montpellier I, 345 p..

Merle, C. (1992). Modèle conceptuel d'évaluation des projets productiques en milieu PMO. Travaux de recherches N° I.9204, IAE-CREGE, University Bordeaux I, 26 p..

Vallespir, B., Doumeingts, G., Zanettin M. (1989). Proposals for an integrated approach to model and design manufacturing systems: the GRAI Integrated Method. *Third International Conference on Computer Applications in Production and Engineering*, Tokyo, Japan, October 2-5, 11 p..

Vallespir, B., Doumeingts, G., Bitton, M., Zanettin M. (1990). GRAI method and economic performance measurement system. *IFIP WG5.3 International Conference on Modeling and Simulation for Optimlzation of Manufacturing Systems Design and Application*, Tempe, Arizona, USA, November 8-10, 1989 - Amsterdam, North Holland, pp. 149 - 170.

Vallespir, B., Chen, D., Zanettin M., Doumeingts, G. (1991). Definition of a CIM architecture within the ESPRIT 'IMPACS' project. *In CAPE'91: The fourth IFIP conference on computer applications in production and engineering*, Bordeaux, France, September 10-12, 1991 - Amsterdam, North Holland, pp. 731 - 738.

THE IMPACT OF A COMPUTER INTEGRATED
BUSINESS ENVIRONMENT ON THE OPERATIONS
OF A MEDIUM SIZED ENGINEERING COMPANY

C R Chatwin, A S Tsiotsias, B F Scott

*University of Glasgow, Department of Mechanical Engineering,
Manufacturing Systems Group, Glasgow, G12 8QQ, Scotland, UK*

Abstract. The adoption of integrated computer based manufacturing and management techniques by small traditional engineering companies often represents an unaffordable and high risk investment strategy in technology that is often not well understood by its recipients. Paradoxically, the opportunity for complete success in a SME is greater than in a large company which very often is incapable of full integration due to the divisions and inertia implicit to a large hierarchical organisation. To derive full benefits from such an investment the company must possess a meticulous understanding of its market, fiscal environment, operations management, engineering and technological skills, manufacturing facilities and product range. It must adopt an appropriate implementation of CIM that does not debase previous *ad hoc* investments in what are often termed islands-of-automation or information technology. For success a well planned stepwise approach is vital. This exposition reports on the approach adopted by a small to medium sized Scottish Engineering Company specialising in the production of mechanical actuation systems. Over a three year period the company embarked on a - low cost - phased implementation of software and hardware systems that exploit a database to integrate its design, manufacturing and business operations. The company's investments were based on a prudent assessment of its current and planned product range, existing and planned manufacturing facilities, the scale of its operations and business objectives. The impact of this technological change on the business operations of the company and its influence on the evolution of its product range are discussed.

Key Words. Computer Integrated Manufacture, Computer Aided Design, Parametric design, feature based design, Computer Aided Manufacture, database integration, integrated business environment.

1. INTRODUCTION

Fortune Engineering Limited is a medium sized Scottish Engineering Company with a turnover of £3m (4.3 MECU) and a workforce of 51 employees; the company designs and manufactures mechanical actuation systems. The main product groups are: precision worm screw jacks; 'Spiracon' ™ planetary roller screws; linear actuators; 'Rolaram' ™ planetary roller screw actuators. Each product group can be manufactured in endless variants of size and power to meet different duties, but, in essence they are conceptually similar. A roller screw, for example, may have variable thread pitches, screw lengths and diameters, nut sizes and mounting arrangements depending on the application and the dynamic load. However, it remains a roller screw.

The product range is sold for a wide variety of applications e.g. precise horizontal alignment of rollers in a steel or paper mill, removal of fuel rods from the core of a nuclear reactor. Fortune Engineering units have been fitted to all the safety doors in the Channel Tunnel and doors on the European Wind Tunnel in Germany; they control the one-tonne rear doors on the Warrior military personnel carrier. One substantial order currently being completed is for the British Airways Dragonfly project at Cardiff, a maintenance facility for wide bodied jets. Fortune actuators power the rigs which will lift entire 747 jets for maintenance purposes. The product range is supplied from a standard catalogue or as special customer designed units, customised units comprise 44% of Fortune's market. Reduction of the time to tender for their special designs was crucial to the business and parametric design techniques were appropriate.

Prior to the initiation of the Teaching Company Scheme with Glasgow University, Fortune Engineering possessed two CNC machine tools which were programmed directly from their Fanuc controllers. The programmes so generated were stored on cassette tape with off-line editing being achieved via an NEC portable computer. The only computing facility was a Data General machine with 5 terminals. This machine was used to run the accounts and MRP system. Tooling for machine tools was under the control of individuals, there being no overall management system. No CAD system existed: all design and design modifications were done on drawing boards; calculations were manual. Production control and sales relied upon a paperwork system. A major debilitating business difficulty with the old manual system was that as the volume of work increased Fortune Engineering would have to take on additional staff which often resulted in less profit being made as costs rose more rapidly than income.

Fortune Engineering's original computational environment was designed to service solely the needs arising from accounting practice and customer order processing. The systems employed were mainly concerned with the costing of finished, catalogued products; manufacturing planning was restricted to the provision of an account of the component parts which had to be delivered by a set date. There was no account detailed of the description of the manufacturing processes to be employed, the flow of materials, components, tooling and fixturing requirements. Furthermore, planning was only feasible in cases where the final product had been manufactured previously. Variations from the 'standard' set were treated as special cases, even if the new item was merely a slight modification of an existing part. The inability of the system to manage multiple variations of a single product led to a high degree of duplication and redundancy of product information.

The Managing Director recognising the need to introduce new management and manufacturing technologies and realising that new recruitment and the development of existing staff were crucial issues, elected to engage the Teaching Company Scheme (TCS). This is a national scheme jointly funded and managed by the Department of Trade and Industry (DTI) and the Science and Engineering Research Council (SERC). The Scheme exists to change traditional attitudes and to form lasting partnerships between higher education and industry in order to improve UK industrial performance, profitability and management. Its aim is to train and develop young engineers to advanced levels in new technologies and secure technology transfer from the HEI sector to companies through collaboration, focussed on company needs and objectives. The TCS provides 70% funding for labour, 50% for capital equipment with the balance being paid by the recipient company.

The objectives of the Fortune Engineering Scheme were :

(i)   to design and install CIM incrementally to existing methods;

(ii)   to cause major reductions in lead times on product variants by introducing parametric design methods;

(iii)   link design and the control of materials, tools and machine tools to secure effective manufacturing cycles;

(iv)   to improve market responsiveness, particularly for overseas markets, by speeding up the tendering process for products which have to be engineered to customer specific requirements. This currently represents 44% of turnover.

These issues were central to the further expansion of Fortune Engineering as they were the sources of vital strictures affecting production. Their resolution was essential to the future profitability and health of the company. The particular approach chosen involves the incremental integration of market/design and design/manufacturing facilities, this being most appropriate to the needs of SMEs.

From the inception it was recognised that the introduction of CAD/CAM would bring cost savings and wider access to markets. Furthermore, investments in CNC machine tools could not be realised until NC code could be generated directly. Over the three year period of the project the company embarked on a low cost, phased introduction of software and hardware systems. A database was used to integrate its design, manufacturing and business operations. The company's investments were based on a prudent assessment of its current and planned product range, existing and planned manufacturing facilities, the scale of its operations and business objectives. The primary achievement at Fortune Engineering is the integration of manufacturing operations control with the design and tendering procedures.

There was an investment in CNC machine tools of £1 million (1.43 MECU), in computer hardware of £100k (143 kECU) and software £50k (71.4 kECU). The communications and control infra-structure that was created allows production volumes to fluctuate with no concomitant increase in the cost of managing the additional throughput. The introduction of CIM has secured these investments and the future well being of the company.

2. SYSTEMS DEVELOPMENT STRATEGY

Accurate classification and description is the basis for coherent engineering operations and allows the integration of design and manufacturing with business functions. A single product register is the initial building block, enabling products to be completely identified and described; furthermore, it provides the opportunity for a company to restructure its software resource and link its manufacturing functions. The drawback of such an approach is that existing software and hardware systems may be made redundant. Fortune's relatively small investment in the technology meant that reorganisation could be managed without significant risk to the company. New systems could be chosen based on specific company requirements and introduced in parallel with the paper processes which they were to replace. The opportunity to modularise the transition process was highly beneficial as it allowed systems to be built and tested in advance of any changes in operational procedures, thus minimising the disruption often encountered.

The development strategy included :

(i)   a vigorous analysis of cost benefits to be derived and of the build of investments and incidence of operational costs;

(ii)   a formal procedure for : analysis of current systems; development of the specification of requirements, its issue and review; approval by the Engineering Systems Department of systematic changes envisioned; consultation with all staff and seminars to secure their aid; company wide announcement and approval of plans for implementation of changes;

(iii)   analysis of system build into modules not exceeding three months in duration, so as to win early advantage from investments and training and for ease of control;

(iv)   in software : the adoption of industry standards and commercial packages; new code and adaptations of application packages required to meet company specified system requirements written and maintained in-house; similarly, in-house creation of manuals and installation and validation of software;

(v)   systems authority - and, therefore, the product description - maintained in the central database with applications holding only temporary copies of data;

(vi)   a plan for staff training and time allocated to it within the overall plan for change;

(vii)   of paramount importance, the exercise of authority by the Managing Director in system analysis, planning and communication with staff.

Whilst the application of this simple set of rules proved to be outstandingly successful for Fortune Engineering, it is inadequate for large companies. Fortune Engineering benefits from being sufficiently small that the Managing Director is still capable of maintaining complete control of his organisation. In larger organisations, where this is not the case, the company management structure should be reviewed to assess what implementation problems may arise. The management review would almost undoubtedly reveal that changes in management structures are necessary before CIM could be successfully adopted.

3. SYSTEMS DESCRIPTION

Fig.(1) is a schematic of the hardware that has been installed. There are 3 Apollo DN3500 Workstations, two of which run Ferranti CAMX 2-D Draughting; the other supports the Ingres Relational Database Management System. There are 16 Compaq 386 PCs which support Ingres applications and custom PC applications. An Ethernet (TCP/IP) network has been installed throughout the factory and forms the backbone of the communications between the design workstations and PCs. The number of CNC machine tools has been increased from two to six.

Fig.(2) illustrates the software platform that supports the CIM environment depicted in Fig.(3). The primary achievement at Fortune Engineering is in the integration of manufacturing operations with design and tendering procedures, a link so often not attained by manufacturing companies. This is illustrated by Fig.(3). A database and its control systems provides the link between what would otherwise be islands of automation. It is rare to find such a high level of integration even within large companies. Fig. (4) illustrates how the system developments are integrated into the business environment. A key issue has been the commitment of the Managing Director who has supported these developments with his time and policies. In larger organisations ambitious plans, as implemented at Fortune Engineering, can founder due to inter-departmental rivalries. Paradoxically, the opportunity for complete success in a SME is greater than in larger companies which are very often incapable of full integration due to the divisions and inertia implicit in a large hierarchical organisation.

FIG.1. HARDWARE CONFIGURATION

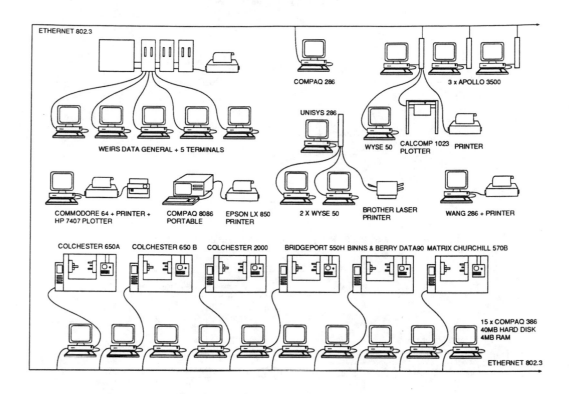

FIG.2. SOFTWARE CONFIGURATION

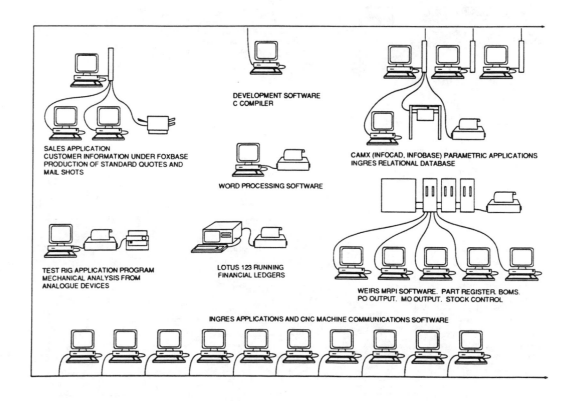

FIG.3. CIM ENVIRONMENT

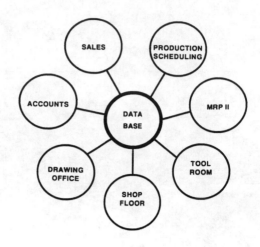

FIG.4. INTEGRATED BUSINESS ENVIRONMENT (IBE)

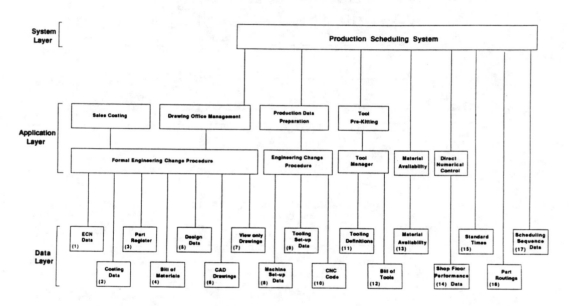

FIG.5. PROJECT 142 - LEAD TIMES USING THE IBE

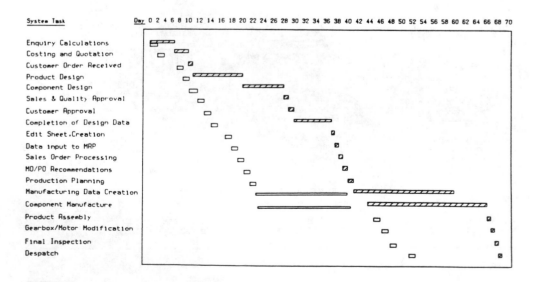

A well integrated three-tier management structure exists at Fortune Engineering; this highly cohesive and controllable structure has allowed the University to make a substantial impact on the company. Regular briefing sessions - with all employees, at all levels of the organisation - on the projects progress and future development have been a major element in the acceptance of new methods and hence the project's success. Training was also a key factor contributing to the success of this project, with relevant employees being given substantial off the job training.

4. MECHANISMS FOR PRODUCT CLASSIFICATION

Product classification is directly associated with design, production and process control, manufacturing resource management and requirements planning. The availability of local expertise in the systems to support design coupled with an immediate company requirement to update and expand its product range suggested that maximum benefit could be derived by, initially, concentrating the company's efforts in the area of CAD. The perceived ability to produce early results in what constituted a well-known bottleneck within the company, would also enable the 'conversion' of personnel associated with manufacturing operations to the idea of a close-coupled, integrated, functionally driven system.

Parametric descriptions of products

A salient characteristic of Fortune's product range is the large number of proprietary calculations, constraints and techniques employed during the design of the individual components. Several computational packages had been developed over a number of years to aid the designers in their task. The inputs to these calculations are often based on customer requirements and, in most instances, were sufficient to describe the products. In the paper-based system access to this information was severely restricted, thus forcing the designers to perform the entire set of calculations whenever a new product was required, resulting in long lead times for development and poor response-to-tender times.

Furthermore, as no complete record of the design was kept, it was impossible to accumulate information relating to manufacturing costs or the preferred techniques to be employed on the shopfloor. Fig.(5) shows the progress of a customer order from its receipt to completion. The requirement existed for a system which would allow information to be gathered whilst the product traversed the different phases of development, and for this information to be represented in a form which could be made readily available to the engineering disciplines for future reference. From a systems viewpoint, this required the provision of a computer based design facility incorporating the existing proprietary software and a storage mechanism to act as the global information repository.

Parametric design addresses the ideas of knowledge encapsulation and representation in the form of interlinked, mutually constraining, hierarchical structures. (e.g. shape-related features on a component together with concomitant information on the tooling and manufacturing processes required to form them) and allows constraints on values and processes. Systems are available which allow for the development of parametric programs to produce geometric representations of objects. Furthermore, schemes have been proposed which enable the use of predefined engineering shapes (e.g. counterbores, slots, chamfers), or features, in the design of products. Parametric geometry facilities are, normally, a pre-requisite for feature-based system implementations.

In the case of Fortune Engineering, the main product descriptors are the most important information items; these must always be retained. Drawings are viewed as a simple manifestation of the component and not as a complete representation of the design record. Feature based modelling is an attractive and elegant solution, provided that the designers are allowed the freedom to define the features and their attributes based on considerations drawn from their own expertise. In addition, the ability to specify explicitly the source of

values for these feature attributes as outputs of specialised software is of paramount importance. Most of the systems commercially available only allowed for the use of predetermined features alone and are of very limited utility. An alternative was to develop a system in-house using the programming facilities available in a normal commercial geometric modeller (InfoCAD).

In addition to the obvious advantages which such an approach entails in terms of user-customizable features, the need to generate a drawing record is eliminated, thus reducing the requirements for the extensive storage capacity normally associated with CAD systems. Drawings can be easily and automatically produced, following a request for viewing, based on the information stored. Information derived from product design calculations can also be stored in a database, closely coupled with the system, and form a product description repository for classification purposes. This is available for consultation, using standard querying techniques (SQL), during the design of new products. Entire assemblies can, therefore, be classified in the system and referenced in future designs thus creating a facility for a group technology description and its support.

Implementation

Parametric design has been successfully introduced for all the main product groups; design lead times have been reduced dramatically from nine to two days. Database structures have been created to hold the geometric definitions of component parts. Programs have been written to create, modify and search for these definitions. This provides a quick means of loading parts with manual drawings on to the CAD system, defining materials and proprietary parts for selection and reusing components in future designs. A generalised definition and drawing method is also being developed to extend the scope of the system to all turned parts, not just the standard families. This Feature Based Design and Manufacture package has substantially reduced lead times.

Control Database

The database structures created to support the activity of product design were significant in that they provided the basis for an initial product classification. The data generated comprised information which could be used to derive product Bills of Materials directly from the database. Generation of part numbers and descriptions is organic to the process of design of a new part and Engineering Change management has been facilitated as a simple, programmed extension of the database update utilities.

The MRP system already implemented at Fortune had been used to perform most of these product control functions. Its inability to be integrated with the parametric design system, coupled with the problems in its operation and poor performance, when compared with the database utilities, led to its replacement by them. More importantly, the database became the nucleus for information which supported a variety of functions in different departments. The capacity to manage products rather than some elements of them (e.g. part number generation) highlighted the role of the central repository and its integrating nature and led to its adoption as the sole controller of information.

It was further realised that the basic classification system incorporated within the control database to associate geometric, material, costing and specification information with parts and products could be extended to include information concerning the product's manufacture.

Manufacturing resource register

The Manufacturing Department at Fortune Engineering were responsible for determining the processes, tooling, fixturing and flow of products on the shopfloor. The process of manufacture was initiated by customer orders and required information from the design department in the form of released parts. This was normally in the form of drawings with incorporated materials information.

The computational facilities in existence were extremely primitive, completely lacking the ability to generate process or production plans. The introduction of the parametric design system in the Design Department, made available the information necessary to drive these applications in a standard format suitable for exploitation.

The particular structuring of the information in the database implicitly favoured links with manufacturing resource items represented in similar terms. Tooling and fixturing may be suitably arranged into families; a specification pertaining to their use by certain manufacturing processes can be attached to them, enabling retrieval of tooling data based on manufacturing process and materials constraints. Furthermore, manufacturing processes may be arranged into groups and a specification can be attached to relate them with shape and geometric aspects of components. In addition, 'parameterised' CNC code fragments can be attached to the manufacturing processes enabling them to subsequently receive geometric parameters directly from the descriptions of the parts in the database. This structured linking of design with manufacture enables diverse operations to be performed with full knowledge of the constraints applying to them.

Computer Aided Manufacture of all components on the CAMX CAD System has been implemented. A Shop Floor Production Data Management System, based on the Engineering Database, has been developed; this comprises three main modules :

(i)   Production Data Management, which monitors and controls :-

        Machine Setup Sheets, Tooling Sheets, CNC Code,

        Tool Management system which defines and controls all CNC Production Tooling,

        Direct Numerical Control (DNC) System which transfers and controls all Production Data to the CNC machines. View only CAD drawings are available over the network,

(ii)   Production Scheduling System; this provides scheduling facilities for the Production Controllers to load manufacturing orders to the CNC machines,

(iii)   CNC Code proving package which will graphically prove CNC Code off-line.

5. CONCLUSION

The three year technology transfer and training programme successfully introduced an appropriate, low cost, CIM environment to Fortune Engineering. The system exploits a central database to integrate its design, manufacturing and business operations into what is more accurately described as an integrated business environment. Fig.(5) illustrates how a typical customised-product lead time, i.e. from enquiry to dispatch, has been reduced from 69 to 53 days. Almost more significantly the average tendering response time has dropped from 10 to 4 days. If an enquiry is given priority a fully designed solution, quotation and proposal can be with the customer the same day. What Fig.(5) does not show is that as the volume of quotations and customised designs increases there is no concomitant increase in labour costs. Thus, profit margins have been enhanced. The speed and quality of data now available to the marketing operation has directly increased the flow of new orders. As a direct result of this project, the company has penetrated the Japanese market, and is the single source of supply for mechanical actuators to Mitsubishi Heavy Industries.

The new integrated business environment has allowed the design and introduction of a new product range, the 'Rolaram' ™ electro-mechanical actuator, which now accounts for 20% of Fortune's turnover. With the design resources previously available this development would have been impossible.

All CNC programmes and tooling data are stored on the central database. Loading of programmes into the machine tools has been reduced from one hour to a few minutes. Tooling is now pre-kitted and presented to the machine operator. These improvements have increased the throughput of work against standard time by 27%. The company now expects to be able to cope with a three-fold increase in business volume without raising staff levels.

This collaboration, only briefly described herein, has transformed Fortune Engineering's business opportunities and secured its medium term future. Unfortunately there will be many SMEs that fail to make this transformation.

6. ACKNOWLEDGMENTS

The Authors would like to thank Dr Alan Watson of the SERC for his valuable contribution to the management of this project. Thanks are also due to Mr G Silk and Mr S Howard the Teaching Company associates and to Mr Alan Fortune, the Managing Director of Fortune Engineering, who had the vision to invest heavily in new technology.

7. BIBLIOGRAPHY

Bell, D., Morrey I. and Pugh J. 1987, *"Software Engineering*, A Programming Approach", Prentice-Hall International (UK), Ltd., London.

Chatwin, C.R. 1991, University of Glasgow - Fortune Engineering Teaching Company Scheme, SERC Final Report, Grant No. GR-F-09631, Reference No. TCS/FR/GLU/CRC910930.

Chittock, B. and Whittington, D. 1990, "MMS and OSI: Key to Manufacturing Communication", Integrated Manufacturing Systems, Vol. 1, October, pp. 205-209.

Gutschke, W. and Mertins, K. 1987, "CIM: Competitive EDGE in Manufacturing", Robotics and Computer-Integrated Manufacturing, Vol.3, No. 1, pp. 77-87.

Heninger, K.L. 1980, "Specifying Software Requirements for Complex Systems. New Techniques and Their Application", IEEE Transactions on Software Engineering, Vol. SE-6, No. 1, pp. 2-13.

Kaminski, Jr., M.A., "Protocol for Communication in the Factory", IEEE Spectrum, April 1986 pp. 56-62.

Merchant, M.E. 1985, "Computer-Integrated Manufacturing as the Basis for the Factory of the Future", Robotics and Computer-Integrated Manufacturing, Vol. 2, No. 2, pp. 89-99.

Rana, S.P. and Taneja, S.K. 1988, "A Distributed Architecture for Automated Manufacturing Systems", The International Journal of Advanced Manufacturing Technology, Vol.3, No. 5, pp. 81-98.

Sommerville, I. 1989, *"Software Engineering"*, Third Edition, Addison-Wesley Publishing Company, Inc.

Tsiotsias, A.S. and Muir, C.D.R. 1992, "Prototyping Tool to Relate Components, Materials, Tools and Processes for Design", CAPE 8 Conference Proceedings, Edinburgh.

Yeh, R.T. and Zave, P. 1980, "Specifying Software Requirements", Proceeding IEEE, Vol. 68, No. 9, pp. 1077-185.

A LOW COST CONTROL STRUCTURE FOR FLEXIBLE PRODUCTION LINES WITH MULTITASK WORKCELLS

Th. Borangiu, A. Hossu

Department of Industrial Process Control, Faculty of Control and Computers, Polytechnical Institute of Bucharest, Romania

Fl. Ionescu

Fachbereich Maschinenbau, Hochschule Konstanz, Germany

Abstract. Flow-type organization of production lines is often encountered in small and medium sized companies. In order to improve the through put, to ensure a high quality of the products and to reduce the manufacturing time, flexible automation is necessary, which natuarlly leads to the CIM approach. The paper describes a flow-type production line for can filling, which can be utilized both in the petrochemical and in the food industries. Considering the typical production line as a numerically controlled workcell, a parallel production structure with multitask workstations is proposed, leading thus to an efficient extension of the company's shop floor. In such a parallel production line (PL) there are a number of identical multitask workcells (MTWC) which include a fixed number of workstations, several Industrial Robots (IR) with interchangeable grippers and tools and diverting devices. The workstations perform all the necessary operations in order to obtain the final products. Thus the productivity of a PL is given by the sum of the productivities of its parallel branches. For such a process structure, the paper proposes a hierarchical control configuration. At the level of local automation, a number of Programmable Logic Controllers (PLC) are used. Product trans fer and handling requires the connection of the PLCs with the controller of two IR, in order to ensure the proper material flow between the production branches. Manufacturing in the PL is integrated by a high-level computer, which analyses order entries, performs job scheduling for the branches which execute identical products, diverts products of different type, downloads programs and reference data to the local controllers, receives information concerning the current state of the production and modifies the job flow according to temporary stops or failures in the individual branches.

Keywords. Flow-type production line; parallel production structure; multitask workcell; flexible automation; programmable logic controller; hierarchical control; computer integrated manufacturing.

INTRODUCTION

A production line (PDL) consists from a number of technologic resources (machines, devices) linked by a transportation system for the material flow.

Automation of PDL provides the attribute of flexibility, with two levels of on-line process control:

- local, with numerical control equipments associated to the machines, devices;

- central, with a hierarchical controller which analyses order entries and schedules the jobs (parts, assemblies, partial processed products) through the line's resources.

Also, application programs are downloaded by the central controller, and automatic diagnosys of the entire PDL is performed by on-line status data acquisition and pre eessing.

In the deterministic scheduling theory with respect to the resources (machines),

a set of jobs (parts, products) must be processed with extremization of a certain cost function. The optimization problem is subject to inequality constraints, which are introduced by the machining parameters (Borangiu, 1991).

According to the way in which the jobs are visiting the resources, there exist two fundamental processing modes, which impose also the layout of the machines:

- flow-type organization all the primary products p_i are circulating in a unique direction between the machunes M_i, as shown in Fig. 1;

- job-type organization, in which the products circulate in any direction.

Each operation e_i can be characterized by the following data (parameters):

- t_{ij} : processing time of operation i on machine j;

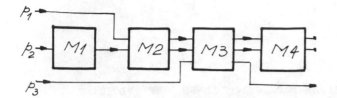

Fig. 1. Flow-type organization of a PDL.

- r_i: availability of operation o_i for execution;

- d_i: imposed time for finishing operation o_i.

For a set of operations, the precedence constraints can be defined with respect to the operations. The relationship $o_i < o_j$ signifies that processing of o_i must be finished before o_j can be started. Thus, the set of operations is ordered by $<$, which is a precedence constraint.

A set of operations ordered by precedence constraints is usually represented as a directional graph in which the nodes correspond to the operations, and the arcs to the precedence constraints.

For each operation o_i, $i = 1, 2, .., n$, the following parameters are defined:

- C_i: time of finishing operation o_i;

- F_i: flow time - the sum of machining and auxiliary times.

$$F_i = C_i - r_i$$

A job scheduling procedure for which the value of such a cost index is minimized is called optimal on-line.

For the following analysis, the mean flow-time will be considered as cost function, subject to optimization

$$F = (1/n) \sum_{i=1}^{n} F_i$$

At the same time, the particular version of the flow-type production line has to be considered for the set-up procedure and on-line scheduling activity of the central controller:

- serial flow, with circulation through the whole chain of machines, which allows, using the aggregating concept, to establish the initial rate of the PDL (Tchijov, 1989);

- serial flow, with circulation of jobs penetrating the chain of resources at different points and possible reentrance.

PARALLEL PRODUCTION LINES WITH MULTITASKING WORKCELLS

A serial flow-type PDL with complete circulation contains a number of workstations each of them performing a single operation upon the product (monotask workcell).

Certain monotask workcells can be equipped with robots, which extend the capacity of material processing from a single task to a macrotask, as a sequence of monotasks with local control of the sequence of operations, speed of material processing, material handling, quality inspection, a.o (Fig. 2)

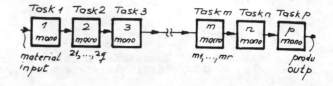

Fig. 2. Complete serial PDL with mono- and macrotask workcells.

Once defined a complete serial PDL with automatic control by means of local numerical equipments, the structure can be easily extended to a parallel production line (PL), which consists from a number of identical workcells (PDLs), as shown in Fig. 3.

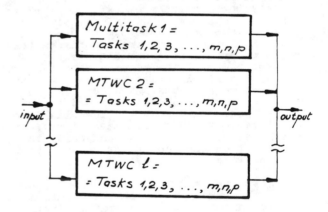

Fig. 3. Parallel production line with multitask workcells.

Such a PL is usually equipped with industrial robots for material handling at the terminal sections of the line raw material feeding and final product diverting and storage.

The productivity of a PL is given by the sum of the productivities of the individual branches.

The multitasking workcells (MTWC) can operate independently; thus, the failure of one branch only reduces the global productivity with the local contribution, but does not stop the entire production structure.

It is obvious that such a PL does not represent a simple concatenation of n multitask workcells, but creates a high-performance complex production structure being supervized by a central computer which performs the following main actions:

- receives entry orders from a hierarchical planning level of the production;

- according to the current state of the individual MTWCs, changes dynamically the program of the robot which feeds with raw materials the production branches, thus scheduling products on-line;

- alters the individual transfer rates of the MTWCs, depending on feedback information received from the automatic macrotask cells and on the current status of the robotized storage station for final products;

- diverts products in case of temporary failures or stops of the individual branches, in order to maintain a full unmanned control of the entire production line, preserving a high throughput of the PL (De Koster, 1988).

PARALLEL MULTITASK WORKCELL LINE FOR CAN FILLING

This production line is dedicated for can filling in various industries with specific petrochemical and food products.

Figure 4 presents the global structure of this parallel production line, and its control resources.

The production line which consists from four production branches (PDLs) runs fully automatic under the control of:

- a central computer which supervises the routing of the empty cans which are downloaded from a central storage by the industrial robot IR1; in consequence, the motion program of this robot is dynamically altered according to the current status of the individual filling branches;

- two robots IR1 and IR2, the second one controlling the downloading of filled cans;

- four PLCs, each of them dedicated to an individual multitask filling branch organized as a complete serial PDL.

Each filling line is organized as a MTWC and consists of five monotask workstations namely : empty can feeding (in pairs); tightness control of cans; can filling; can plugging; transfer line evacuation.

The structure of the PDL for can filling, identical for the four branches in the PL configuration, is given in Fig. 5.

A conveyor (transfer device) is displaced in steps by a cylinder - piston (A) mechanism at a rate given by the initial set-up

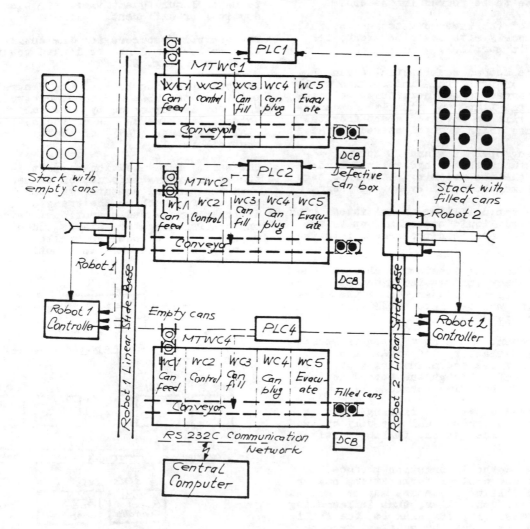

Fig. 4. Parallel MTWC line for can filling

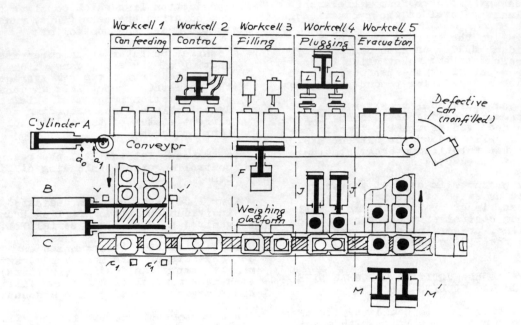

Fig. 5. Structure of PDL for can filling.

of the line and the filling parameters: quantity of liquid to fill the can, number components to be poured in the can.

Each one of the five workstations operates simultaneously with the other four, for two cans in parallel.

If, at the second workstation, a can is detected nontight, this information will be transferred to the next three workstations of the line, and finally the defective can will be diverted apart from the filled ones which are palletized by IR2.

The filling operating and PL control sequence assumes that a hydraulic cylinder F pushes the two weighing platforms, each of them being occupied by an empty can.

Then, the rapid filling is initiated, with the selected liquid component, up to the weight which has been programmed into the PLC.

The rapid filling is stopped when the numeric feedback information given by an incremental linear encoder equals the reference value written into the PLC by the central computer.

At this moment the slow filling is started which takes place by deactivating one of the feed devices and maintaining the liquid flow only through the second device, for a prescribed time period Δt.

As mentioned above, the filling cycle is duplicated and runs simultaneously on two identical devices at the filling multitask cell.

According to the information produced at the tightness control workstation, one or even both filling processes may be skipped by the logic controller, thus influenceing the PDL rate. (If, for example, the filling station requires the greatest processing time).

Finally, the filled cans are downloaded from the five PDLs and palletized in a

simple stack, or in dedicated stacks, depending on the fact that the can sizes, forms or liquid quantities, types are identical or different.

The control resources for the entire parallel production line are linked together as shown in Fig. 6.

The communication between the central computer and the PLCs and robot controllers RC1, RC2 is supported by a serial RS 232C network which makes use of the on-line serial interfaces (OLSI) of the six local controllers.

The serial comminication support is used for an intensive information exchange between the central computer and the local controllers, such as the transfer of appli cation programs and operational data from the central computer, status and diagnosys report from the local controllers, off-line programming of the two robots.

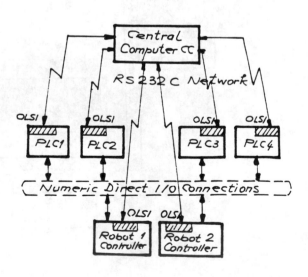

Fig. 6. PL control network

22

The operational interconditioning and synchronization between the multitask work cells are obtained by means of direct multichannel connection of the local controllers via the numeric I O interfaces; this techiique is used for the rapide communication between RC1 and PLC1, .., PLC4, respectively between PLC1, .., PLC4 and RC2.

CONCLUSIONS

An efficient automation solution for complete serial production lines is possible both by considering the production line as a multitask workcell, and by using Industrial Robots at the terminal areas of the parallel extension of a number of PDL branches.

For each of the process branches, the transfer rates are first established during system analysis by means of the generalized aggregation procedure (Tchijov, 1989 and Meerkov, 1990).

Then, an efficient communication network is built, providing both rapide operational interconditioning between the controllers, and also the possibility to modify dynamically the MTWC's rates and the handling programs for the terminal robotized areas.

REFERENCES

Borangiu, Th. and R. Dobrescu (1986). Programmable Controllers, Academic Press, Bucharest, Romania, 212 pp.

Borangiu, Th. and A. Hossu (1991). Analysis and Synthesis of Flexible Manufacturing Systems, Polytechnic Press, Bucharest, 137 pp.

De Koster, M. B. M. (1988). An Improved Algorithm to Approximate the Behaviour of Flow Lines, Int. J. Proc. Res., 26, 137-142.

Meerkov, S.M. and F. Top (1990). Asymptotically Reliable Serial Production Lines, Synthesis and A Case Study. Proc. of the 11th IFAC World Congress, Vol.9 Tallin, Estonia.

Tchijov, I. and A. Alabian (1989). Flexible Manufacturing Systems, Main Economic Features. Proc. of the 2nd IIASA Ann. Workshop on CIM, Laxenburg, Austria.

Order related- and personnel time acquisition
as part of a CIM/OCA-solution

M. Zauner*, P. Kopacek* ** and G. Steinegger***

Dept. of Systems Engineering and Automation, Scientific Academy of Lower Austria, Krems

** *Institute of Robotics, University of Technology, Vienna, Austria*

*** *KEBA Company, Linz, Austria*

Abstract: The company under investigation is divided into three strategical units (profit center´s): "bank-, sawmill- and industry automation". On the other hand of the matrix-organisation of this company there exist the sections marketing, finance and system-technology. It is intended to find an operating characteristics acquisation system (OCA) which is embedded in a computer integrated manufacturing (CIM) solution. To reach this goal the project will be realised in two steps, first of all to introduce the personnel time acquisition and afterwards to develop the order related data acquisition system. The concept of this OCA solution has to take into account several frame conditions. On the one hand an existing LAN has to be used to transport the OCA data. On the other hand the data management will be realised by means of a relational database management system under SQL, in order to obtain an open interface to ensure an expandability of the data base for the second level and the integration of the order related data acquisition in the non production areas. The complexity of the project should be reduced by using standard software packages in separated modules.

Keywords: operating characteristics acquisition, relational database, client/server software architecture

1. Introduction

Since the company was founded, the acquisition of personnel and order related data was done by using a time stamp. It is one and a half year ago, that the demand occured to develop a concept for an operating characteristic acquisition (OCA). On the one hand this demand was caused by the state of technology, which offers several possibilities to automate data acquisition. On the other hand the demand increases to unload the employees from manual data recording and at the same time reduce costs which result from wrong data inputs.

The organisation of the company is devided into three profit centers, a bank-, sawmill- and industry automation unit. The internal organisation is devided in several sections, for example marketing, finance, production and technology. A main part of the incremental value development in this company are the domains of product design, construction and software-engineering. In the production area the prints for the machine control system are complemented both, in surface mounted technology (SMT) and manual. The software-programms in the EPROM´s are self developed and implemented.

Due to these different fields of work the personnel data acquisition system must map many operating time modells. The order related operating data acquisition system must input data in the company´s PPS system, which has to administrate the orders of all profit centers.

2. Frame conditions

2.1 Organisational conditions

Operating data acquisition systems involve all domains of a company. The matrix-organisational structure of this company increases the complexity of the project, because there are three virtuell sub-companies, which describe their special requests to the OCA solution. The goal is, to support the sequence of operations in all domains of the company very well.

The working hours for the employees are very different. In the company there exist both, a flexible operating time, for example in the domain of design and software-engineering, and a rigid operating time in the domain of production.

The project is devided in

1) personnel time acquisition and

2) order related data acquisition in the

2a) non production areas,

2b) production area.

The application of step 1 is placed in the commercial section, where costs of labour are computed. At this point interfaces to the domain of cost accounting have to be taken into consideration. The project was realised under the head of the section systems-engineering of the company in cooperation with the instituts of the co-authors.

An important restriction was to install standard software packages to reduce the complexity of the project.

It was decided to realise step one in 1991 till Juli 1992, and the first part of step two till december 1992. The second part of step two will be realised in 1993 till 1994 because for this step many requests occure from the PPS system, which is already reconditioned at the same time.

2.2 Technical conditions

There is a local area network, which connects all buildings of the company. This FDDI based network, called the backbone, supports ethernet and the TCP/IP protocoll. Inside the buildings, there are several sub-networks, which are connectet to the backbone-network via bridges. The workstations of the sub-networks are connected with coaxial cables. The CIM data-base server is a HP-9000/730 workstation. In addition the commercial domain uses an IBM-AS400 mini computer to run their applications. Because operating data are also used in both, the technical and the commercial domain, an interface between the RDBMS-Oracle on the HP9000 and the AS400 data base (file structure) has to be taken into account.

The physical infra-structure between and inside the buildings must be utilized to transport the OCA data from the input-terminal to the data-base server. It was forbidden to lay cables between the buildings.

The data management will be realised by means of a relational database management system under SQL, in order to obtain an open interface to ensure an expandability of the database for the second level and the integration of the order-related data acquisition in the non production areas and afterwards in the production areas.

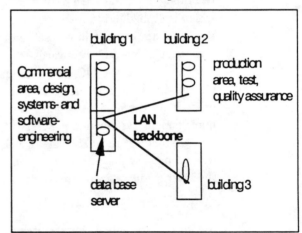

Figure: network infra structure

3. Concept

The concept was devided in two steps. The first one was to specify all detail characteristics of the software applications. For example, two of these characteristics of the personnel time computing system are the description of operating time models and the description of kinds of cost of labour.

The second step was to design a concept that takes into consideration all organisational and technical restrictions.

Therefore the existing relational data-base management system Oracle was defined to be the central data base of the whole OCA solution.

It was not possible to connect the OCA-terminals to the LAN directly, because in 1991 there was no company that could offer such a solution. All firms worked hardly to solve this problem, but at this time no firm has ended the problem solving process.

To hold the project´s deadline it was decided to find an alternative way to connect the OCA terminals to the network. A main restriction, the forbiddenness to lay cables between the building was still upright.

An alternative solution for the problem was to connect an OCA-concentrator between the OCA-terminals and the network. The advantage of this alternative was, that the data transmission between the buildings could be realised with the use of the backbone network (as it was defined at the beginning of the project). The agreement was to lay cables between the OCA terminals and the concentrator inside the buildings. The additional wiring inside the buildings is based on a seriel RS485 network.

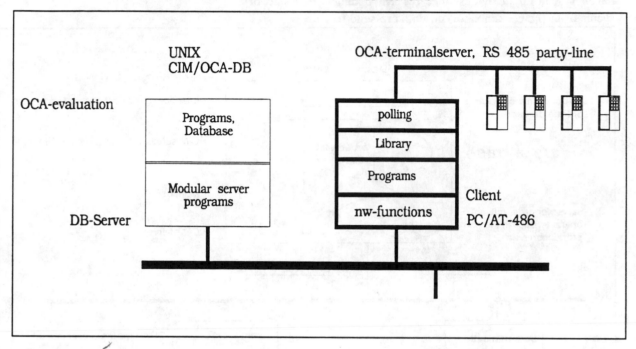

Figure: The software architecture of the OCA solution.

To feed the application with the order related times in the non production area as well as software-engineering, the existing PC´s and application terminals are used. All data are administrated in the common OCA data base.

Due to the use of OCA concentrators the software design has taken into consideration the advantages of a client/server principle.

With this concept a common relational data base is used to administrate the OCA data. These are the personnel data and in the following step the order related data in the non production area. The database server is also the application server for the personnel time computing software package. Therefore a great advantage is that Oracle is also used as the personnel time computing software package´s database.

The server-programms on the database server communicate with the OCA concentrator via an interface file. This interface file is placed in a special directory on the database server, too.

So, all messages which are determined for the database, and in the next step for the software package, are fed into this interface file from the OCA concentrator. For example, those messages are all activities, that are stamped with the OCA terminal.

In the other way messages from the software package are also put in this interface file. These messages are determined for the OCA terminals. Therefore the concentrator has to read these messages from the interface file on the database server and inputs them into the OCA terminals. For example, such a message is generated, when an identifaction card is lost.

The concentrator is polling the OCA terminals. It has to be taken into consideration that all collected data have to be send to the database server immediately. This request is due to the three different buildings in the company. There are special groups of employees, who have to work in all buildings and not only on one working place. For these employees the personnel time stamps must be load in all OCA terminals immediately.

To minimize the network utilization the concentrator only connects to the database server, when the OCA terminal has send data. If the OCA terminal is "empty", that means it haven´t been any activities since the last polling cyclus from the concentrator, no data are send to

itself. In this way the concentrator is intelligent to know, that no connection to the database server is nessecary.

The concentrator is a PC AT-486, which is configured as a network bridge. Therefore the concentrator transforms the messages that arrive from the RS-485 party-line in a low level protocol into a high level protocol. With this format the messages are send to the HP9000/730 database server with the use of the ethernet backbone network.

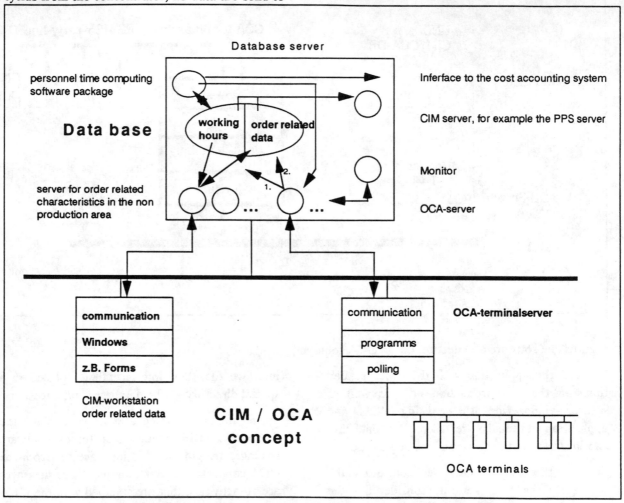

Figure: The OCA concept

4. The instructional strategy

When starting the project a course was shaped for the sequence of operations. First of all the requests and restrictions have described. Some of the requests have not been formulated detailly, because it was planned to take into account the ideas of the hard- and software suppliers.

The sequence of operation to select a software product run in the opposite way as it is done

usually. In this project the restrictions in the hardware components and infra structure have forced to select the software packages of the suppliers in the following way:

- is it possible to integrate the existing LAN infra structure?

- no cable laying between the buildings.

- does the software support a SQL interface ?

- is Oracle as database supported ?

- can all operating time models and the company's special problems to compute the personnel working hours be mapped by the software?

- minimized network utilization

You see, there have been very differents requests which were due to both, the demand of the user to operate with an ideal software package and on the other hand the domain of systems-engineering, which has planned to develop a companies wide common database and therefore a uniform hardware platform has to be used to avoid compatibility problems. So it was decided to order the software by one supplier, that has to look for all hardware and software components.

Because of the different request it was inpossible to find a supplier who could solve the problem with the use of standard moduls in 1991.

There were three suppliers who tried to install their solutions, but no solution has taken into consideration all requests. In addition to this fact there occured essential troubles and errors in the application during the test phase, so it could not be found any agreement to one of this suppliers.

It was a great luck that the fourth supplier could solve the problem, because the deadline of the project was past over six months and this situation has to be justified against the head of the company.

A very important step in the instructional strategy was to introduce an in-house test phase before signing the contract. Due to this fact there have been avoided many troubles and costs, because the user and the system administrator of the company could test the system in the desired infra structur and check all functions. When all tests have been all right the contract was signed.

5. Conclusion and Experiences

One experience was that many software suppliers sell their software packages without proving the requests of the companies. So it is worth to define with the supplier to test the whole software package from competent employees. It is to recommend to generate a test enviroment inside the company under "life-conditions".

This kind of test must be insert into the instructional strategy when buying standard software packages and before a contract is closed.

Today, a OCA solution is running, which uses the relational data base system Oralce to administrate all operating characteristics. The communication between the OCA terminals and the database supports the client/server concept with the use of an OCA concentrator. This concentrator operates as brigde between the ethernet network and the party-line network. The intelligence of the concentrator is used to minimize the utilisation of the network. The order related data of the non production area will be fed to the data base via client-applications under Oracle Forms. This applications will run on the CIM-PC's of the company and via VT-100 emulation from many other terminals.

6. Literature

Kopacek P., Frotschnig A., Zauner M.: "CIM for small companies", Automatic Control for Quality and Productivity - ACQP '92, IFAC Workshop, Proceedings, Istanbul 1992.

Zauner M., Kopacek P.: "Data integration in CIM concepts for small and medium sized companies", Computer Aided Technologies - CAT '92, User Congress, Proceedings, Stuttgart 1992.

PC-BASED HIERARCHICAL MANUFACTURING CELL CONTROL

E. Freund, H.-J. Buxbaum, U. van der Valk

Institute of Robotics Research (IRF)
University of Dortmund
Otto-Hahn-Strasse 8, W-4600 Dortmund 50, F. R. Germany

Abstract. Robots in the industrial factory plant build the backbone for CIM, which is outlined to
be the factory automation strategy of the future. In this paper a hierarchical manufacturing cell
control system for robotic work cells is described. This approach bases on low cost personal
computer hardware and consists of two different control systems, one for the robot control tasks
and another one for the work cell control tasks. The underlying concept is based on a strategic
architectural model for hierarchical manufacturing control, where it focuses on the process
sequence coordination and the process control in the bottom hierarchical layers. The cell controller coordinates the components in the work cell and offers an open interface to control and planning systems beyond cell level. Full production flexibility by individual product identification as
well as universal cell configurability are characteristic for the cell controller system. In the manufacturing control hierarchy the robot controller serves as a connecting link between the cell controller and the robot hardware. It executes predefined robot programs and on demand informs the
cell controller about its current status. The integrated menu oriented programming environment
of the robot controller provides an efficient tool for implementing the robot programs. The
IRDATA interface of the controller permits the execution of programs, which have been generated by external programming systems with a corresponding interface.

Keywords. CAM; Flexible Manufacturing System; Work Cell Control; Robot Control; PC-based
Controllers; Low-Cost Automation

INTRODUCTION

The major applications of robots are production plants for
large batches with very small product variations. The
actual trend to small batches, large product variaty and
far-reaching considerations to individual customer requests shows the necessity to develop new factory
automation strategies with the aim to enlarge the production flexibility. A migration from normal machine control
technology to a comprehensive factory wide planning,
coordination and control structure is indicated. The robot
itself is a suitable device for the purpose of flexible
automation, if only because of its kinematic structure.

Small and medium sized companies have special requirements for their factory automation. Due to small
batches and high flexibility they need automation equipment, which supports their approval to individual pur-
chaser requests. Customized controller solutions are often
very expensive which makes it very risky to start with
automation efforts. In this approach the work cell and
robot controller software is implemented in a PC-envi-
ronment, which makes it a low cost solution, suitable for
many different applications.

Basing on a universal work cell controller conception
(Buxbaum, 1990; Freund, 1992) here a personal computer implementation of the cell controller software is
described. The task of this system is the coordination of
the different automation devices in a flexible manufacturing work cell. The controller conception is based on a
hierarchically structured architectural model for factory
control. Production flexibility is achieved by individual
product identification. The production process for each
unique product is programmable by the operator. The
process sequence in the work cell is explicitly described
by a product specific process plan. Since the work cell
controller is implemented under MS-Windows, all operator activities are full window- and mouse supported. An
open communication interface to the factory floor components for DNC allow universal work cell configurability.

Parallel to the cell controller a PC-based robot controller
was developed. The main aspects behind this development were
- to implement a complete robot controller on the low-
  cost personal computer hardware,
- to design and implement an integrated menu oriented

environment as an efficient and easy-to-learn tool for program generation,
- to provide a powerful interface for execution of robot programs constructed by external programming systems and
- to offer an interface, which allows the integration of the robot controller in complex automation hierarchies.

In realizing the above objects the PC-based robot controller is capable to serve as a flexible, low-cost automation component in the manufacturing control hierarchy.

HIERARCHICAL MANUFACTURING CONTROL

Flexible factory automation requires a manufacturing control system for the planning of manufacturing activities and for controlling of the production equipment in the factory floor. It is the aim to ensure that the products ordered by the customer can be delivered on schedule. Fig. 1 shows the desired transparency of the information flow from the customer order through the manufacturing control system to the production equipment on the factory floor. The feedback from each point is included. Besides controlling the information flow, the manufacturing control system also provides coordination mechanisms for the internal flow of material and goods.

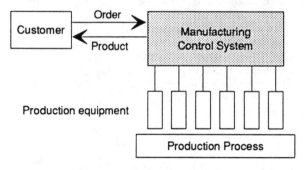

Fig. 1: Organization of manufacturing control

Fig. 2 shows a 5 layer architectural model for manufacturing control. Each hierarchical layer is assigned tasks from the layer above and in turn delegates sub-tasks to the layer below. The 5 layers cover the full range from production and factory systems (top) down to the sensors and actuators in the factory floor (bottom). The layers also extend over various time horizons which encompass both long-term planning and real-time processing. The

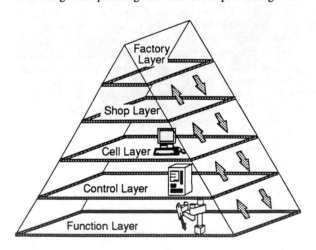

Fig. 2: Architectural model for manufacturing control

essential meaning of the model coincides with comparable approaches by Albus (1981) and Pritschow (1990).

STRUCTURE OF THE FLEXIBLE WORK CELL

A work cell is a cooperating machinery unit in the factory floor which carries out entire manufacturing operations autonomously. The objective of those operations is to bring a product from one defined state of manufacturing to the next. A work cell usually consists of a few automation devices such as robots, machines and transport systems (production equipment).

In flexible systems this equipment is used for families of products. The variants of the product families can be manufactured by means of programming or data transfer. For each product variant an individual production plan has to be prepared in the preliminary stages of manufacturing. The production plan describes all production steps to be done on the product.

The process activities within the flexible work cell can be devided to the three fundamental functions *product identification*, *internal logistics* and *production process*. The *product identification* must be done everytime a new product arrives in the work cell. The tasks of *internal logistics* are the transport and the storage of products and parts within the work cell. The *production process* is described by the production plan and appropriately delegated to the production facilities, which are often robots for reasons of flexibility. Fig. 3 shows a characteristic control hierarchy in a flexible manufacturing work cell and the proper fundamental functionality of each branch.

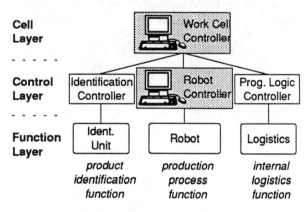

Fig. 3: Hierarchical layers in the flexible work cell

An exemplary layout of a flexible work cell is given in fig. 4. At the import point all products, which are reaching the work cell, are identified. Via a node in the transport system the product can be transported either to the workplace or directly to the export point via the bypass structure. After being processed by the robot the products leave the work cell.

PC-BASED WORK CELL CONTROLLER

The heterogenous control structure on the factory floor and the demand to universal applicability leads to a coordination concept based on a library of specific device drivers. For each device in the control layer of the work cell configuration a specific driver has to be installed. A

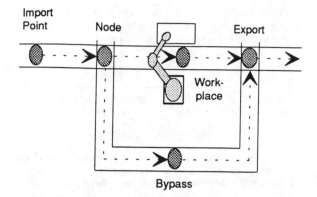

Fig. 4: Exemplary layout of a flexible work cell

defined communication protocol between those device drivers and the coordinator kernel allows to standardize already on the level of the coordinator's application software and control interfaces. Fig. 5 shows the input mask on the operator screen for the definition of a new control layer module. The operator defines a name for the new module and selects one of the available device drivers.

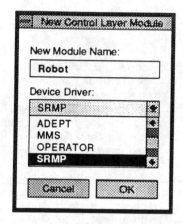

Fig. 5: Control layer module definition mask

During runtime the work cell controller coordinates the control layer modules according to process plans. The demand for flexibility makes an assignment of an arriving product to the corresponding process plan inevitable. This assignment is done by product identification at the very beginning of the manufacturing process for each unique product. The rules for this assignment are provided by the workorders. A workorder specifies the identification range of products which belong to this order and includes a reference to a process plan. Fig. 6 shows the input mask for a workorder on the operator screen. The workorder links the appropriate process plan to a range of product identification numbers.

The process plan points out all process steps to be done on a product in the flexible work cell. All non-productive steps like material flow and error handling are included in the process plan additionally. Related to the process plan in each step a program is started on the associated device controller. After the ready message is received, the execution process continues with the next step. Decisions depending on the execution result of the previous step allow the handling of error states as well as intelligent processing.

The number of different products which can be handled by the work cell simultaneously is not limited by the cell

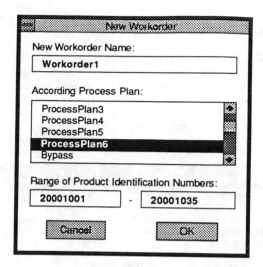

Fig.6: Workorder definition mask

controller. Nethertheless there are physical limitations like the number of work places and buffers in a real work cell.

The representation of the process plan - the sequence of the concerned automation devices and corresponding parameters for the device controllers - can be realized in different forms. The description in a schematic graphical form will be called process scheme, the expression process plan stands for the state table representation of the process sequence. For a batch of unique products - as a single variation of the well known product family - a particular process plan (or scheme) is needed.

Fig. 7 shows a process scheme for the examplary work cell (fig. 4). After product identification on the control layer module *Transport* the machine program *Node* is performed. This results here in transporting the product to the node. The product is transported to *Workplace* in the next step. *Robot* performs its program *NameOfProg* on the product before it leaves the work cell in the last step. In case of a robot error during execution an *Opera-*

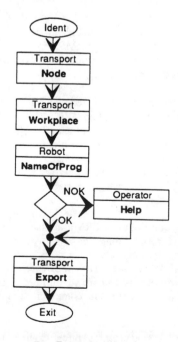

Fig. 7: Process scheme

tor call is generated before the product leaves the work cell.

Fig. 8 shows a screen hardcopy of the work cell controller's process plan editor, where the process plan according to this process scheme can be seen. The process plan is the internal data representation of a process scheme for use by the cell controller. Each row contains the information to perform a process step in the work cell. The first two columns contain the process step number and

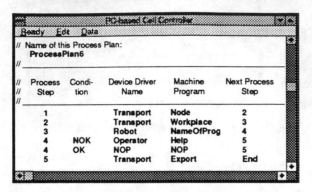

Fig. 8: Process plan editor

the result status of the previous step, so that conditional branches through the process plan are possible. The next two colums contain the device driver name and the machine program to be executed there for this step. The last column references the counter number of the next process step.

COMMUNICATIONS

The communication between the cell controller and the robot controller is done by the use of a defined DNC interface. This interface is subject of standardization. ISO 9506 proposes the network technology MAP and the communication protocol MMS as standards for communication in a manufacturing environment with the goal to ease the link of automation systems to an information structure.

Kerndlmaier (1991) defined the Simple Robot Management Protocol (SRMP). SRMP provides application oriented communication services for robot controllers. In contrast to the more extensive approach of MMS only DNC-functions for the link between cell controller and robot controller are supported. SRMP is based on a communication model and an ASN.1 (ISO 8824) protocol definition similar to MMS, but uses TCP/IP as the underlying transport protocol. Any computer system (personal computer, workstation) providing the widely available TCP/IP-Socket-Programming-Interface can be interfaced to SRMP. Cheap and easy-to-install Ethernet can be used as network technology.

The service definitions of SRMP are based on an abstract model of a robot controller ("Virtual Robot Control VRC"). The services are a functional subset of corresponding MMS services. They, however, are reduced in scope and complexity and are streamlined to the needs of robotics. Fig. 9 shows the communication services of SRMP.

The general management services contain the Initiate,

Conclude, Cancel and Status services. They allow the user to initiate and to conclude communication with another SRMP-user, to cancel pending service requests and to determine the general status of the connected SRMP-server. The domain management services contain

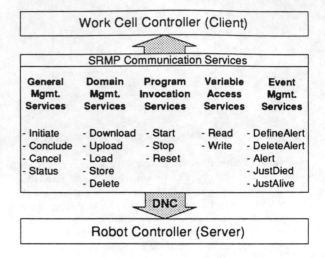

Fig. 9: DNC-interface with SRMP communication services

the Download and Upload services for program transfer between the client SRMP-user and the server. The Load, Store and Delete services support the internal program management functions of the SRMP-server. The program invocation services provide facilities (Start, Stop and Reset) which allow the client SRMP-user to start, stop and reset the actual program of the server. The variable access services Read and Write allow the SRMP-client to access variables (such as calibration status, actual joint values, speed factor) defined at the server. The event management services provide facilities (DefineAlert, DleteAlert, Alert) which allow the SRMP-client to define and manage event objects and to obtain notifications of event occurrences. The SRMP-server can also announce sudden events (JustDied, JustAlive).

PC-BASED ROBOT CONTROLLER

In the control hierarchy of a flexible work cell the robot controller serves as a connecting link between the cell controller and the robot hardware (Fig. 3). The main task of the robot controller in the work cell is the execution of predefined robot programs on demand of the cell controller. To be able to carry out its process plans correctly the cell controller requires information about the current status of the robot controller. From this another important task of the robot system in the cell hierarchy results. To be capable of performing the above tasks the robot controller should provide
- an appropriate hardware and software interface to the cell controller,
- an efficient tool for implementing and testing robot programs and
- a well defined interface for the execution of programs, that were developed using other robot programming systems.

Following a controller is described, that fullfils the above requirements. The PC-based robot controller was designed and realized as a low-cost system for robots of

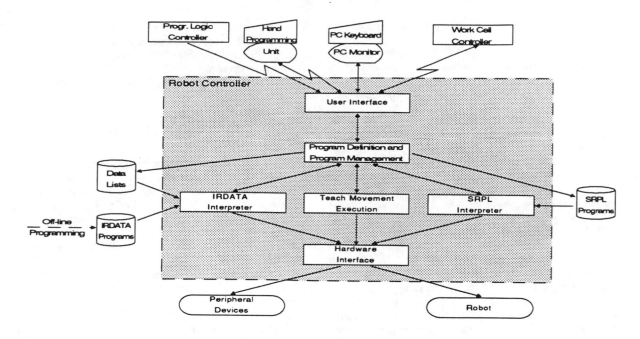

Fig. 10: Basic structure of the PC-based robot controller

the lower price level. Additionally one main aspect behind this development was to implement an integrated menu oriented programming environment as an efficient and easy-to-learn tool for program generation, which can also be used by novices in the field of robotics. Especially this aspect seems to be substantial for the intended application areas, which are handling and assembly tasks, education and research and laboratory applications.

The hardware basis of the robot controller consists of a personal computer. The PC is additionally equipped with special-purpose hardware, that performs the axis control for the robot axes, provides an interface to the servo amplifiers and supplies the robot controller with digital input and output channels. The operating system basis of the controller is MSDOS, the standard operating system of the PC-world. Especially no real time kernel and no multitasking extensions are used. Integrated in the controller is a complete on-line movement control with the movement types synchronous point-to-point, linear and circular.

The most important decision in designing the robot controller was the selection of the PC as the hardware basis. The main reasons for this decision were:
- The PC-hardware including the operating system MSDOS is available for much lower costs compared to other hardware platforms intended for controlling tasks with real-time and multitasking operating systems.
- The system prize can even be more reduced, if a PC is utilized as control computer, that is already available in the place, where the robot controller is installed. The significance of this aspect bases on the large distribution of PCs. Especially this reason is of interest for the application area of education.
- In idle times, when the robot is not employed, the controlling PC can be used for other purposes refering to the professional and well known tools of the MSDOS environment.

MODES OF OPERATION OF THE ROBOT CONTROLLER

The basic logical structure of the PC-based robot controller is shown in the data flow plan in Fig. 10. It is best explained by a description of the different modes of operation of the controller (Fig. 11).

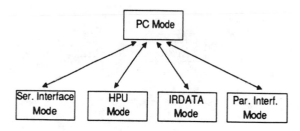

Fig. 11: Modes of operation

The main operating mode of the robot controller is the so called *PC mode*. In this mode the user operates the controller by keyboard and mouse. Messages generated by the controller are displayed on the monitor. After initialization of the controller the user is in this mode. Only if working in the *PC mode* the user can switch to another mode. In the *PC mode* the user operates in an integrated menu oriented programming environment. The menus are implemented as pop-up and pull-down menus. Mainly the following functionality is realized in this mode:
- program management,
- program generation,
- program execution,
- teaching of the robot.

As the basis for the programming system a simple robot programming language *SRPL (Simple Robot Programming Language)* was defined according to the main aspects, that
- programs in this language can be generated efficiently in a menu oriented environment,

- the operation of the system is easy-to-learn even for non-experts in robotics and
- the functionality covers most of the problem solutions in the intended application areas.

In the *HPU mode* the robot controller is operated by a hand programming unit. The functionality of this mode corresponds mainly to the functionality of the *PC mode*, although the realization is adapted according to the capabilities of the hand programming device. In both modes the SRPL programs are executed by the SRPL interpreter. An additional feature of the *HPU mode* is, that it provides an interface for the definition of data lists by the user of the robot controller. Data lists are special files, from which and to which IRDATA programs read and write data (especially robot position data).

The *IRDATA mode* provides a powerful interface to external programming systems. A serious problem in the industrial application of robot systems depends on the fact, that robot programs and taught positions are generally not interchangeable between different robot controllers, because each of these uses its own programming system with a special programming language. Based on these considerations a standardization proposal for a low-level robot programming interface was designed by a working group of the Association of German Engineers (VDI). This proposal, *IRDATA (Industrial Robot DATA, DIN 66313)*, is intended as a standard interface language between a robot programming system and a robot controller (Fig. 12).

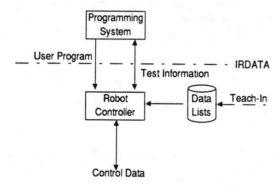

Fig. 12: IRDATA as an interface language

A robot controller with an IRDATA interface is capable of executing IRDATA programs. Therefore the interchangeability of programs, namely IRDATA programs, between robot controllers with an IRDATA interface is guaranteed. Additionally the integration of the data list concept in IRDATA provides the possibility to interchange taught positions between different robot controllers. With the IRDATA interpreter the PC-based robot controller is capable of executing complex robot programs, which have been generated off-line by external programming systems.

In the *Serial Interface mode* the robot controller is operated by a higher level automation unit using the serial interface of the personal computer. In this mode the robot controller acts as a server, that receives commands (such as loading of a program, starting and stopping the execution of a program, opening of the gripper) and sends replies (status and error messages). As an extension of the *Serial Interface mode* a complete DNC-interface based on SRMP is in preparation.

In the *Parallel Interface mode* the robot controller is operated by a programmable logic controller using digital input channels of the robot system. In this mode only elementary commands for loading of a program and starting and stopping the execution of a program are defined.

CONCLUSION

A hierarchical manufacturing control concept was presented, which consists of a work cell controller as the coordination system in the cell layer and a robot controller as the control layer system. Full production flexibility via free programmability of the process sequence in conjunction with individual product identification on the one hand and universal cell configurability via integration of device specific driver software on the other hand are characteristic for the work cell controller. The main aspects in realizing the robot controller were to implement a menu oriented programming environment as an easy-to-use tool for program definition, to integrate an interface for execution of robot programs defined by external programming systems and to provide the possibility to link the controller to higher level automation units.

One of the main features of the system components is their cost effectiveness. In this approach personal computers are used as hardware platform exclusively. The conception is independent of the process and the product and therefore adaptable to many fields of application. These are substantial prerequisites especially for small and medium sized companies to start with factory automation efforts.

REFERENCES

Albus, J. B., Barbera, A. J., Nagel, R. N. (1981). Theory and practice of hierarchical control. *National Bureau of Standards*, Washington DC.

Buxbaum, H.-J., Hidde, A. R. (1990). Flexible Zellensteuerung - Bestandteil eines produktunabhängigen Fabrikautomatisierungskonzepts. *Werkstattstechnik 80*, 133-136, 262-264.

DIN 66313 (1989). IRDATA: Schnittstelle zwischen Programmiersystem und Robotersteuerung. Allgemeiner Aufbau, Satztypen und Übertragung. *Deutsches Institut für Normung (DIN)*, Berlin.

Freund, E., Buxbaum, H.-J. (1992). Universal Work Cell Controller for Flexible Automation with Robots. *Symp. on Intelligent Components and Instruments for Control Applications*, Malaga.

Kerndlmaier (1991). SRMP - Anwendungsorientierte Kommunikationsdienste für Robotersteuerungen. *Technischer Bericht, Institut für Roboterforschung*, Dortmund.

Pritschow, G. (1990). Automation technology - On the way to an open system architecture. *Robotics and Computer Integrated Manufacturing*, Vol. 7, No. 1/2, 103-111.

A CAPP SYSTEM FOR
AUTOMATIC TOOL SELECTION IN DRILLING OPERATIONS

J. ISOTALO and K. JANSSON

VTT, Technical Research Centre of Finland, P.O. BOX 111, SF–02151 ESPOO, Finland

Abstract. The paper presents the process planning of prismatic workpieces which are usually manufactured in a machining centre. The main interest has been laid on an automatic tool selection based on the information of tools used previously in similar manufacturing operations. A Computer Aided Process Planning System for Machining Centre Operations has been developed.

Key Words. Process planning, tool selection, feature modelling, drilling.

INTRODUCTION

Today, industrial production is determined by both decreasing batch sizes, increasing product diversification, high quality requirements and shortening product life cycles. Shorter design to product lead time provides for the competitive edge.

In many companies the production is increasingly customer oriented. As the life cycle of the product is shortening continuously, companies all over the world try to reduce the time to market of their new products. The lot sizes are decreasing and even one-of-a-kind production will be soon everyday work. The lead time from scratch to ready product is or will be dramatically reduced. This trend imposes great demands both on the product design, the production design and on process planning activities.

The share of the costs of the process planning activities in the overall costs is steadily increasing. While in the 1980's the lot sizes were still thousands of units in one batch, in the 1990's the companies must be able to manufacture several hundred different products, each with a batch size of ten units, and all at lower costs and with shorter delivery times. This challenge can be met only by means of automation, information technology and flexibility in the production. CAPP-systems promise assistance and automation of the manual tasks [LAWLER *et al.*, 1990]. It can be seen that the ultimate aim of CAPP in a CIM environment is to take the description of a part, apply planning logic and then produce the manufacturing instructions without human intervention. Naturally, within

a period, the transition to this new technology will provoke a number of problems to both the production planning and the process planning.

The process planning does require much information, knowledge and experience. The knowledge is partly based on experience gained from manufacturing similar products previously. This expertise is stored in a number of different media. Some expertise on the engineering work of a product is not directly accessible as it may exist only in the mind of an expert. Thus, for process planning activities, the information has to be evaluated and collected in some way.

At present, the integration of CAD and CAM is not yet completed. However, by means of Computer Aided Process Planning (CAPP) integrating the product design with the part manufacturing functions can be progressed. CAPP is a substantial step towards Computer Integrated Manufacturing, CIM. CAPP has the potential to provide the essential link between CAD and CAM in order to approach the goals of CIM [ALTING, 1986; BOK & NEE, 1988; CHU & WANG, 1988; SUTTON, 1989; and WANG & WYSK, 1988]. It could be said that feature based modelling of manufactured products and their manufacturing processes is now regarded as a key technology in achieving high levels of efficiency, automation and integration in CAD and CAM. Feature based design, process planning and manufacturing systems are for this reason nowadays widely investigated in all industrialized nations.

The lack of manufacturing information in the product model produced in the design phase has caused some problems both for process planning and for the process planning automation. For the integration of design and other manufacturing activities this had been a major obstacle. Therefore feature modelling has greatly increased the information content of the product model, which is now more capable to act as an input for process planning.

PROCESS PLANNING

The task of process planning is to generate all plans that are necessary to transform a part from raw material to the finished part. A process plan therefore represents the description of the schedule of all individual manufacturing steps and the allocation of working stocks [HUMM et al., 1991] The main elements of process planning are

❑ selecting the processes and sequencing,

❑ selecting the machine tool and the tools,

❑ defining the machining steps and the sequences related to these steps,

❑ considering dimensions and tolerances,

❑ selecting operational parameters.

All elements can be integrated into one system.

Conventional CAPP-systems, even the most sophisticated, seldom do produce more than a limited workstation routing and the bill of materials [LAWLER et al., 1990] as there lacks in most cases a comprehensive feature-model for the part data that can be used for CAPP purposes.

PROCESS PLANNING IN SMALL BATCH PRODUCTION

In Finnish manufacturing industries the production is often a *small batch* production or even *one-of-a-kind*. In these cases much more process planning operations are required than in serial production.

There are some differences in the stage of process planning between small batch production and serial production: in small batch production any capable working process plan solution is often fully adequate. Whereas the flexibility of the manufacturing system is of utmost importance, the manufacturing process itself usually is not carried out as effective as it is done in mass production. As occasionally the process planning in small batch production is considered as a bottleneck, this project was fo-

cused on the development of process planning operations.

NEED FOR AUTOMATION OF PROCESS PLANNING

Modern information processing technology, especially knowledge based techniques, offers the possibility to take a closer look at automation of process planning activities. In one-of-a-kind production and in small batch production there is a great need for automating the time and resource consuming process planning phase. The lead time from design to manufacturing can be reduced by automating the process planning. Also the designer needs manufacturing support when he is making decisions related to part design and manufacturing. Thus there is a need in the design phase for this kind of tools that are aiding the design process or designer in the process of decision making.

PROCESS PLANNING AUTOMATION

Process planning is widely considered to be the act of creating a routing sheet and time standard information [LAWLER et al., 1990] This kind of view of the process planning function makes it difficult for many manufacturers to justify the investment in automation of process planning.

The automation of process planning has met some problems on its way. HUMM et al. (1991) refer the following difficulties

❑ Lack of complete part representation. The product model must contain beyond the geometrical information also technological information on different levels of abstraction to support different planning tasks.

❑ Process planning is characterized by strong interactions between subtasks that may be resolved successfully by human experts.

❑ Up to now, no coherent body of theory has been developed for process planning.

❑ Given the dynamic nature of manufacturing technology, the system must support updating of knowledge to guarantee plans with high quality over a longer period.

One of the main tasks of process planning is the selection of tools. For years engineers have been strongly involved in analyzing the tool selection and in developing tool selection strategies. Cutting tools are selected according to the gathered knowledge from the manufacturing site.

In machining centres, holes and hole making are the major geometrical elements (features) or operations that are carried out. The precision and tolerancing of the features make the hole-making to a demanding task area from the point of view of machining operations and tool selection.

The tool selection principle is based on the definitions of the features given by the designer. The machining methods, operations and operation sequences should be selected so that the requirements of a given feature can be accomplished. For these methods and operations the cutting tools are then selected so, that the given measures of the feature can be attained. The selection process as such is based on the manufacturing expertise from the manufacturing site as found in the knowledge base of the system. The information content of this knowledge base can be extended any time by the user. This allows to *standardize* the operations carried out in the production on the manufacturing cell level. When some of the skilled personnel leave the company, their knowledge remains in the system and the enterprise can carry on with its production without the need of time consuming teaching periods for the new personnel.

SANII & DAVIS (1990) draw the same conclusions: *The CAPP-systems will maintain valuable process planning capabilities and experience when planners retire or leave the company.*

WORKELEMENTS

Usually a process plan is composed of small units each performing a single small task. These tasks are called workelements. One workelement contains the work done in a single machining operation by one tool.

In an example the drilling of a hole with a specified surface quality and position tolerance requires the utilization of three different tools. Then three distinct workelements are needed to produce the desired hole (*workelement chain*).

Additionally, a process plan can contain elements that are called **auxiliary workelements**. These workelements are usually needed in the process plan but do not involve a machining operation. Examples for auxiliary workelement are

❑ the change of a tool,
❑ the rotation of the pallet, or
❑ a positioning operation.

FEATURE MODEL

The geometry of the workpieces and final parts or products should be stored in a product model. Preferably feature modelling techniques should be used.

One basic problem in the data transmission between MANUFACTURING and DESIGN is, how to explain to the designer what is feasible in the production and what not. This problem can be solved by using predefined features for product modelling. The design is carried out by using features that are defined and agreed by the manufacturing and design function. Thus, information is passed in both directions.

The integration of CAD and CAM requires new methods for product data definition, communication and management. The modern CAD-systems are most often based on graphics. Originally they had been designed for drafting, drawing processing, editing and filing. Technical drawings do contain all information. So all phases following CAD can be based on that information.

Process planning, like most of the other computer-aided engineering applications, requires as input the definition of the geometry of the product. However, information on the geometry alone is not sufficient as complete product definition as process planning reasoning requires the engineering significance of the geometry to be captured as well.

The concept of features has been introduced by researchers that attempted to create links between design and manufacturing. The motivation was the desire to generate process plans and NC-programs partially or total automatically directly from the design definitions. Today, many definitions have been offered in the literature: "*Features are generic shapes with which engineers associate certain properties or attributes and knowledge useful in reasoning about the product*" [SHAH, 1991] or "*a feature is a region of interest within a part or — more generally — within a product*" [NIEMINEN et al., 1991]. It must be mentioned that the feature definition is dependent on the viewpoint. For example, for a designer, features are elements used in generating, analyzing or evaluating designs whereas from the manufacturing viewpoint, features represent shapes and technological attributes associated with manufacturing operations and tools.

According to SHAH (1991) the three basic approaches in which features can be used are

❑ interactive feature definition,
❑ automatic feature recognition,
❑ design by features.

The features can be divided into different types:

form features:
elements related to nominal geometry

precision features:
acceptable deviations from nominal form or size

technological features:
performance parameters, etc.

material features:
material composition, treatment, etc.

assembly features:
part relative orientation, fits, interaction surfaces.

Rather often *features* and *attributes* are used as synonyms. But, whereas an **attribute** is a characteristic or quality of a thing, a **feature** is a physical constituent of a part having en engineering significance and predictable properties [SHAH; 1991]. Attributes can be applied at any level to a feature, or a collection of features, to a whole part, or to an assembly. Attributes may include several pieces of information, e.g., mating surfaces, fits, material specifications, part numbers, administrative data, dimensions, shape and size tolerances, geometric constraints, surface roughness and form tolerances.

THE TOOL SELECTION SYSTEM

At VTT a project has been carried out in order to develop an Automatic Tool Selection (ATS)-system [ISOTALO *et al.*, 1992] that creates machining phases (workelement chains) and also selects cutting tools and machining parameters for almost all machining centre operations. The overall description of the system is shown in Figure 1.

The ATS-system uses as input the manufactured features, manufacturing knowledge, cutting tool and material and machine tool data. The final output consists of lists on the selected cutting tools and the machining parameters and the sequenced NC-program for manufacturing. This is performed automatically, but the user is allowed to modify the selection.

The ATS is currently able to handle the following machining operations:

- centre drilling
- twist drilling
- U-drilling
- broaching
- reaming
- boring
- milling (sink milling, bottom milling)

- tapping
- roller-burnishing
- bevel-milling
- bevel-drilling

Currently ATS is oriented in hole making and the following features can be handled:

- bevel
- through hole
- blind hole
- flat bottom blind hole
- stepped hole
- tapped hole

The feature attributes are:

- diameter of the feature,
- depth of the feature,
- diameter tolerance value,
- IT-grade,
- surface roughness,
- Ra-value,
- position,
- position tolerance.

The system output is

- cutting tools,
- machining parameters,
- sequenced workelement for the set-up.

As input for the NC-generator the sequenced workelements are used. They carry all information needed for the generation of the NC-programs. The outcome is the NC-code for the set-up.

The ATS contains the following program modules

- Feature Editor
- Knowledge Editor
- User Interface for the Tool Selection
- Machining Parameters
- Sequencer
- NC-Generator

Their mutual relations are shown in Figure 1.

ATS IN DETAIL

ATS works in four steps. At first workelement chain alternatives for each feature are searched and then the tools are attached for each workelement in the chain. Subsequently the chains are ranked and the highest ranks are selected. Then the selected workelements are sequenced and the tools are allocated on the machining centre.

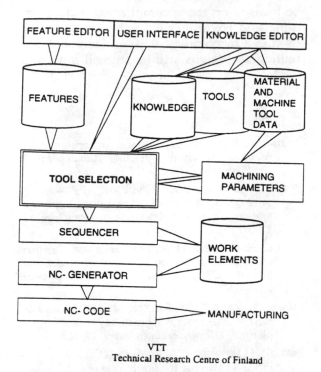

VTT
Technical Research Centre of Finland

Figure 1   The ATS–system.

The user puts feature information into the file. Then the ATS selects machining phases, cutting tools and machining parameters according to the knowledge collected in the knowledge base. Cutting operations are sequenced in proper order and finally the NCprogram is generated.

MANUFACTURING KNOWLEDGE

The knowledge base contains rules and workelement classes. Whereas a rule controls when a specific workelement should be used to manufacture a single feature, a workelement class controls how the manufacturing is performed. Rules do use the concept of workelement chains.

A **workelement** is a basic operation. A set of workelements (containing at least one member) is called a **workelement chain**. For example, *thread* involves both a *drilling* and a *tapping* operation. Thus the workelements *drilling* and *tapping* belong to the workelement chain for *thread*.

A **rule** consists of three parts: *type* of the rule, *left hand side* of the rule (LHS, condition part) and *right hand side* of the rule (RHS, action part). The type of the rule describes on what type of feature the rule can be applied. The rule can be addressed to all feature types that are recognized by ATS, for example, *through hole* or *blind hole*.

WORKELEMENT CHAIN SEARCH

Workelement chain alternatives are found by scanning through the rules. During the scan the following procedure is performed on every rule: the variable *'feature'* is bound to features whose type matches the type of the rule. Then the LHS of the rule is evaluated and if the result is **T** (or **non–NIL**) the workelement chain expressed by the RHS will be attached to the feature.

For example, consider the workelement class of sink-milling, where the values of *tool-dmin* and *tool-dmax* are 10 and 15 mm, respectively. As end mills with a diameter of 10 and 12 mm can be found in the tool library, two instances are created. Here, both the tool cutting parameters and the movement of the tools will be different.

TOOL SEARCH

For each workelement suitable tools are searched. The searching algorithm is depending on the type of the workelement. For example, bevel-milling tools are searched in a different way than tools used for drilling. The behaviour of the algorithm depends on the information in the workelement parameters. For drilling through-holes a longer drill is required than for a flat bottom blind hole with the same nominal depth. If no tool could be found, the chain will be marked and has no associated workelements in the tool selection table. ATS tries to explain why the search was unsuccessful. Possible reasons may be insufficient tool selection control data or that the specified tool does not exist in the tool library.

SCORING THE CHAINS

Once the workelement chains have been found, they will be ranked. The chain with the highest rank will be used to manufacture the set-up.

At the moment ATS-system is based on four ranking criteria, namely

❑ cutting time,
❑ chip removal rate,
❑ if the diameter of the tool is variable, and
❑ if the tool already belongs to the set-up.

The user can redefine the weighting of each criterion after having evaluated its importance for the particular task. By this different strategies can be supported.

SEQUENCING WORKELEMENTS

The sequencing of the workelements is based on the types of the workelements. The order of the groups of workelements is:

1. centre-drilling
2. drilling, U-drilling
3. bottom-milling, sink-milling
4. reaming, broaching, boring, roller-burnishing
5. tapping
6. bevel-drilling, bevel-milling

Even within a group the workelements have to be sequenced. Except for group 3 the sequencing order is determined by the diameter of the tool. For group 3 the sequencing of the workelements is more complicated:

❏ When the workelements belong to the same chain, the order within the chain is preserved,

❏ otherwise workelements attached to a feature with a larger diameter will precede a workelement attached to a feature with a smaller diameter.

SUMMARY

The CAPP-system can create machining phases (workelements), select cutting tools and machining parameters, sequence the operations and generate the NC-program for the machine tool.

One of the principal ideas of the ATS was to formalize the knowledge and expertise filed in the knowledge base. The tool selection is based on this knowledge. The information in the knowledge base can be expanded at any time by the user. This allows that operations carried out in the production can be standardized on the manufacturing cell level. When personnel leave the company, their knowledge remain in the system and the enterprise can carry on its production without time consuming teaching period.

The major advantage of this kind of CAPP-system appears to be in small batch and one-of-a-kind production where the proportion of the process planning is large compared with mass production. In this way the DESIGN and MANUFACTURING operations will be integrated more closely to each other and the dialogue between these activities will increase. This results in shorter delivery times and it will decrease costs.

The follow-up of the system development will be focused on more complicated feature handling and covering all machining operations used in a machining centre environment. Also milling operations will be studied more thoroughly.

REFERENCES

ALTING, L., 1986. Integration of engineering functions/disciplines in CIM. *Annals of the CIRP* **35**:1, 317–20.

BOK, S.W. & NEE, A.Y.C., 1988. MICAPP — A micro computer based process planning system. *Journal of Metalworking Technology* **17**, 21–31.

CHU, C.-H. & WANG, H.-P., 1988. The use of artificial intelligence in process planning. *International Journal of Operations and Production Management* **8**:1, 4–16.

HUMM, B., SCHULZ, CH., RADTKE, M. & WARNECKE, G., 1991. A system for case-based process planning. *Computers in Industry* **17**, 169–80.

ISOTALO, J., JANSSON, K. & PESONEN, O., 1992. Process planning for machining centre operations. *VTT Research Notes.* To be published.

LAWLER, B.D., ROGERS, E.C. & AMSTER, R.B., 1990. Generative Process Planning from PDES: RAMP goes beyond the routing. Proceedings on the Conference *Sheet. Autofact '90..* Detroit 1990 . pp. 16-9 — 16-27.

MADURAI, S. S. & LIN, L., 1992. Rule-Based Automatic Part Feature Extraction and Recognition from CAD Data. *Computers Ind. Engng* **22**:1, 49–62.

NIEMINEN, J., HÄMMERLE, E. & BOCHNIK, H., 1991. MCOES-machining cell operators expert system, Deliverable WP4–D3 (R): Specification of the feature model. Brite/Euram Research project BE-3528. Commission of the European Communities. 1991.

SANII, E.T. & DAVIS, R.E., 1990. Feature-based distributed computer aided process planning system. In *Advances in Integrated Product Design and Manufacturing,* (ed. P.H. Cohen & S.B. Joshi).
American Society of Mechanical Engineers, **PED-Vol. 47,** 163–79.

SHAH, J.J., 1991. Features technology. *Postgraduate seminar of the Institute of Industrial Automation,* Espoo 1991–10–15 ... 16, Helsinki University of Technology.

SUTTON, G.P., 1989. Survey of process planning practices and needs. *Journal of Manufacturing Systems* **8**:1, 69–71.

WANG, H.P. & WYSK, R.A., 1988. AIMSI: a prelude to a new generation of integrated CADCAM systems. *International Journal of Production Research* **26**:1, 119–31.

AN ALGORITHM FOR CAD-BASED GENERATION OF COLLISION-FREE PATHS FOR ROBOTIC MANIPULATORS

G.CONTE, S.LONGHI, R.ZULLI

Dipartimento di Elettronica ed Automatica

Università di Ancona

Italy

Abstract. The use of a CAD system can improve the overall efficiency of automated factories. We describe the utilization of a CAD system devoted to the off-line generation of collision-free paths for robotic manipulators. We first apply this tool in the case of a robotic manipulator in a moderately cluttered environment, next we utilize it to plan collision-free paths for two robotic manipulators sharing a common workspace. The implemented off-line programming system is written in C, in X/Windows environment. In this way portability is effectively achieved. Simulation results show the effectiveness of the proposed approach .

Key Words. Robotics; path planning; collision detection; configuration space; distance field generation; dual-arm motion planning.

1. INTRODUCTION

Companies using reprogrammable devices, such as assembly devices, automated guided vehicles and robotic manipulators, need to stop the productive cycles whenever programming different tasks for the automated devices is necessary . These interrupts of productive cycles increase the equipment down-time and reduce the manufacturing efficiency thus leading to an increase of the manufacturing costs. Therefore, in order to increase the overall efficiency of automated factories, it is important to minimize such interrupts. This problem is still more important for small and medium sized companies where the changes of the program tasks for the automated devices are more frequent. A solution of such a problem can be achieved using a CAD [1] system equipped with simulation and off-line programming tools. Off-line programming allows robot to remain on-line performing manufacturing tasks, while being programmed for another job. Robot simulation and off-line programming require in particular an efficient algorithm for path-planning. This paper describes our research effort in the area of path-planning for robotic manipulators, in particular we focus on the problem of motion planning with collision avoidance. This problem, usually known in robotics as the find-path problem, is described as follows [2]:

- let R be a robotic manipulator, described by k degrees of freedom (joint variables);

- suppose that R is free to move in a three-dimensional space Z amidst a collection of obstacles whose geometry is known to the manipulator;

- given an initial placement S (start position) and a desired target placement G (goal position) of R,

determine an obstacle-avoiding motion of R from S to G.

As a first step for the solution of this problem, we use the key concept of configuration space (Cspace) described in [3]. The configuration of a manipulator is a set of independent parameters, called joint variables, that characterize the position of every point of the manipulator. The robotic manipulator in the Cspace is described by a single point. The configurations forbidden to the robot, due to the presence of the obstacles, can be represented as a region in the Cspace, called the configuration space obstacles(CO). The complementary part of the CO with respect to the Cspace is called the free configuration space (FP [2]). As the Cspace is a discrete space we first state conditions for a feasible choice of the discretization step and, basing on this choice, we calculate the CO. The proposed solution [4] for the calculations of the CO is general, in particular there are no limits for the number of the joint variables and for the shape of the obstacles. Having obtained the CO we reduce the problem of finding an avoiding obstacles path for the robot in the cartesian space to the simpler problem of finding a path for a point in the free configuration space. Many algorithms [2] have been proposed that search for a path in FP. Here we develop a procedure based on a distance-field [5]. The construction of the distance field consists in labeling each point of the FP with a number representing the distance, measured on the minimum length path, between that point and the start position. We begin the construction of the distance-field from the start position: this allows an arbitrary change of the goal position, because distance-field values depend from the choice of the field origin. Using the distance field, we find the shortest path joining S and G in FP simply starting from the goal position and looking adjacent points for the minimum value of the field.

The content of the paper is as follows.

In Section 2 we describe the approach used for the CO calculation and we state the conditions for a feasible choice of the discretization step. The efficiency of the solution and the improvements due to parallelization are also discussed.

Section 3 deals with our solution to path-planning problem in the FP; we describe in details the method used for the construction of the distance field and we exhibit shortcuts and problems.

In Section 4 some examples of application of the proposed method are shown. First we plan a path for a three-dimensional three-degrees of freedom manipulator acting in a moderately cluttered environment. A further application deals with two robotic manipulators acting in a common workspace. The path is calculated in the common Cspace of the two robotic manipulators and the resulting length of the path is minimal with respect to the variations of the joint variables. Simulations show the effectiveness of the proposed solution. The efficiency of the solution and the improvements due to parallelization are also discussed.

Section 5 discuss results, improvements and further applications of the presented method .

2. AUTOMATIC GENERATION OF COLLISION-FREE SPACE

In this section we describe the approach used for the CO calculations following closely the method described in [4]. As the Cspace is a discrete space we first state conditions for the choice of a feasible discretization step.

We use a stick-figure approximation of the manipulator's links and consequently enlarge the obstacles. In the actual implementation we approximate the links of the manipulator with cylinders: then the obstacles are grown in all the directions of the link radius; the links are shrunk down to line segments.

We now state conditions under which a discretization doesn't cause loss of informations.

In fig.1 we see the manipulator's link in two different positions. We need to assure that, given two adjacent collision-free positions of the robotic manipulator, it is possible to move from the first one to the second one without colliding with the obstacles. We assure this fact simply assuming that $P=P_r$ (fig.1).

Elementary calculations lead to the following condition for α:

$$\alpha = \arcsin (2k/1+k^2) \qquad (1)$$

where $k=l/d$ and l and d are respectively the length and the radius of the link.

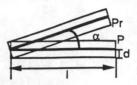

fig. 1
Condition of swept area equal to zero.

Choosing α as in (1) we are guarantied that admissible (collision-free) configurations in the real world are the same obtained from the CO calculations for the grown obstacles and the reduced robot links.

In fact, let us suppose to have two admissible configurations for the reduced link: suppose also that we have a collision in the transition between the two admissible configurations. This implies that there is a point P of the grown obstacle (fig. 2) which belongs to the swept area of the link: this in turn implies, for the choice of the growing factor for obstacles, that there is a collision (fig. 2) in one of the two configurations of the robotic arm. The last statement is in contrast with the hypothesis because we suppose the configurations admissible: thus the claim follows.

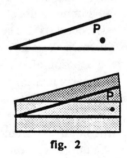

fig. 2

On this ground we choose discretization steps for joint variables: we call $d\theta_i$ the step for the i-th joint variable (in the case of a manipulator composed of only one link $d\theta_i=\alpha$) and $\Delta\theta_i$ the range of variation for θ_i (that is $\theta_i^{max}-\theta_i^{min}$).

We tessellate the surface of the obstacles with triangular patches. A triangular patch has a simple analytic expression. Each point of a triangular patch of vertices $\underline{v}_1,\underline{v}_2,\underline{v}_3$ is represented as:

$$CC(\underline{v}_1,\underline{v}_2,\underline{v}_3) = \sum_{i=1}^{3} s_i \, \underline{v}_i \text{ where}$$

$$0 \le s_i \le 1 \,, \, \underline{v}_i \in R^3 \quad (i=1,2,3) \,, \, \sum_{i=1}^{3} s_i = 1 \,.$$

For the robotic manipulator we have:

$$V(link) = \sum_{i=1}^{3} t_i \, \underline{e}_i + \underline{p} \text{ where}$$

$$\underline{e}_i \in R^3 \ (i=1,2,3), \ 0 \le t_1 \le 1, \ \underline{p} \in R^3,$$

$fl1(t_1)<t_2<fu1(t_1)$, $fl2(t_1,t_2)<t_3<fu2(t_1,t_2)$,

and for a monodimensional link ($\underline{e}_2=0$, $\underline{e}_3=0$);

$V(\text{link}) = t_1\underline{e}_1 + \underline{p}$ where
$\underline{e}_1, \underline{p} \in R^3$, $0 \leq t_1 \leq 1$.

The intersection condition between a robot's link and a triangular patch gives:

$$\begin{bmatrix} \underline{v}_1 & \underline{v}_2 & \underline{v}_3 & \underline{e}_1 \\ 1 & 1 & 1 & 0 \end{bmatrix} \begin{bmatrix} \underline{s} \\ t \end{bmatrix} = \begin{bmatrix} \underline{p} \\ 1 \end{bmatrix} \text{ where}$$

$\underline{e}_1 \in R^3, \underline{v}_i \in R^3$ (i=1,2,3), $\underline{s} \in R^3$, $\underline{p} \in R^3$, $t \in R$.

The solutions of this linear system included between zero and one represent the situation of collision between a link and a triangular patch.

The complexity of the algorithm for the CO calculations is, in the n-dimensional case, $4nm(\prod_{i=1}^{n} \frac{\Delta\theta_i}{d\theta_i})$, where n is the number of the joint variables, m is the number of triangular patches necessary to tessellate the surface of the obstacles, $\Delta\theta_i$ and $d\theta_i$ are respectively the range of variation and the discrete step of the i-th joint variable.

The algorithm is well suited for a parallel execution, since there is no intrinsic serial part. The speed-up for a parallel execution is thus $\dfrac{1}{\frac{1}{p}+\omega(p)}$ where p is the number of processors and $\omega(p)$ is the overhead due to the parallel execution (note that if we neglect the overhead the speed-up becomes p).

Finally, we remark the generality of the proposed solution: there aren't limits for the number of joint variables and for the shape of the obstacles.

3. PATH PLANNING THROUGH DISTANCE-FIELD GENERATION

We now use a distance-field to search for a collision-free path in Cspace. Our construction of distance field is similar to that of [5]. The idea is to label each point of FP with a number representing the distance of minimum length path between that point and the start position. We begin the labeling from the start position: this allows an arbitrary change of the goal position, because distance-field values depend from the choice of field origin. Assuming that FP=Cspace, the construction of this field is equivalent, in the 2-d case, to label each point with his distance value from S calculated by mean of a digital metric, such as chess-board or city-block distance [6].

The choice of city-block distance means that we allow only one joint variable to vary at each step: with this

metric the number of adjacent points is minimum (in the n-dimensional case, 2n).
We call the part of FP which surrounds S inizialized zone.
The algorithm which construct the distance field is described as follows:

1. Inizialization
   - start position has distance value zero ;
   - initially we must consider the adjacent points of the start;
   - initialize L, list of points subject to expansion;
2. Loop: while L is not empty (we calculate distance field for all the points in FP)
   - the point subject to expansion is the first in L (P=head of L);
   Loop1; for each P_i in FP neighbour of P *
   - assign a distance value to P_i
   - store P_i in L (L=[L | P_i]);
   end Loop1;
   - pull out P from L(L = L - P);
   end Loop.
* the number of executions of Loop1 depends from the metric we choose

The complexity of the stated algorithm is $O(\prod_{i=1}^{n} \frac{\Delta\theta_i}{d\theta_i})$.

The stated algorithm has an intrinsic serial part: we can think to this part as the propagation of the distance field. At each step we need the distance values previously calculated. However, the algorithm is well suited for a parallel execution: the operations inside Loop and Loop1 can be executed in parallel. The potential speed-up for a parallel execution is thus equal to p, number of processors. During the distance field construction it is not necessary to consider all the neighbours of P: in the 2-d case, using city-block distance, it is possible to show (**fig.3**) that the number of neighbours we need to consider is less than 4.

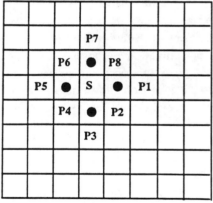

fig.3

The number of neighbours we need to consider is 3 for the points P1, P3, P5, P7 (corner points): we need to consider only 2 points for P2, P4, P6, P8. Corner points are less than the others, so the number of executions of Loop1 is either 2 (more frequently) or three.

The path from S to any point G is now simply calculated starting from G and choosing the minimum value neighbour as next point of the path. We can also change S inside the initialized zone without the need to calculate again distance-field. If we need to find only a path from S to a fixed G, we can stop distance-field construction when we reach G for the first time, thus having a faster execution.

It is also possible to generate two distance fields, beginning rispectively from the start and from the goal position.

In this way it is possible to find a new path for an arbitrary change of start and goal position in FP: the found path is not the shortest. An arbitrary change of start and goal position is equivalent to a motion of Cspace obstacles. We need further calculations only when obstacles' motions change the connectivity properties of FP.

However, to apply the described technique to mobile obstacles, we need to calculate quickly the modification of FP due to obstacles' motion: in this way we can try to modify the path found in the static case in order to obtain a collision-free path with mobile obstacles.

The used method is a complete one, in that we always find a solution when a solution exists.

4. SIMULATION RESULTS

Our simulations were made on an HP 9000/Series 350 machine with a 68020 processor.

A first number of examples is generated considering only one manipulator.

In the 3-dimensional case, supposed to have the three degrees of freedom manipulator and the obstacles of fig.4, the total time for the generation of a collision-free path was of about 7'. Five minutes were required for CO calculations (choosing $d\theta_i = 5°$ (i=1,2,3)) and the remaining two for distance-field calculations.

The found path, being the shortest, is generally close to the obstacles. The reported times of execution are suitable for simulation purposes. Note also that in a problem of gross motion planning as the one we are considering it is reasonable to neglect the degrees of freedom of the end-effector as we have done.

In the sequel we present examples concerning two manipulators sharing a common workspace.

First we present an example concerning two planar arms, each one composed of one planar link. This example, although very simple, is interesting because it clearly explains the relations between the Cspace motion of the point representing the two robots and the real synchronized motion of the two robots. The interesting thing here is that we plan a purely geometric path for the two robots in Cspace but someway the temporal information is retained by mean of the precedent history of the joint variables.

The geometric characteristics of the two robots are described in **fig.5**.

The joint variables θ_1 and θ_2 representing respectively the rotation of OA and BC in the plane with O and C fixed, are allowed to vary respectively between (+90°,-90°) and (+90°,270°).

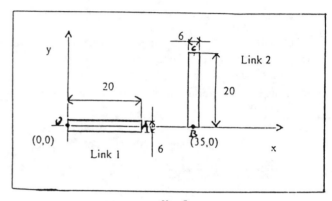

fig.5

Geometric data of the two one-link planar arms.

We choose the discretization step according to (1), obtaining in this case a value of 17.06°.

We plan a motion from the start position S(-60°, 180°) to the goal position G(40°,180°). The found path is represented in **fig.6** with regard to Cspace representation.

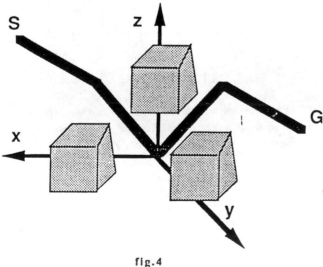

fig.4

A three-dimensional example.

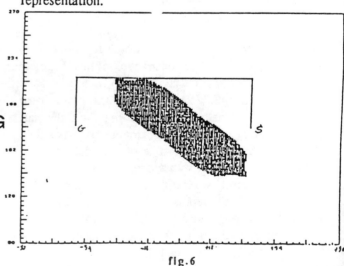

fig.6

Path from S to G in Cspace.

In **fig.7** we represent top-down, left to right the same motion of fig.6 in the real world. The strategy of the motion is simple and clever: Robot 2 must move in order to yeld Robot 1, till Robot 1 achieves his goal position; finally Robot 2 come back to his original position that is also, in this case, the goal position.

Finally we consider two manipulators, each one having two degrees of freedom, moving in the plane with an obstacle.

Geometric data for the two manipulators are summarized in **fig.8**. The start position is S=(-90°,-90°,270°,270°) and the goal configuration is G=(90°,90°,180°,180°).

The found path is reported in **fig.9**. In this case the discretization step is always calculated by mean of (1), considering that the worst case is when the two links of the robot are aligned: the resulting value for α is 4.24 degrees. Execution time in this case, with a four dimensional Cspace, is of about 15 minutes, including the time needed for CO calculations. Note also that the execution time for a three dimensional example for two degrees of freedom arms is essentially the same, because the major factor of growth of execution time is Cspace dimensionality.

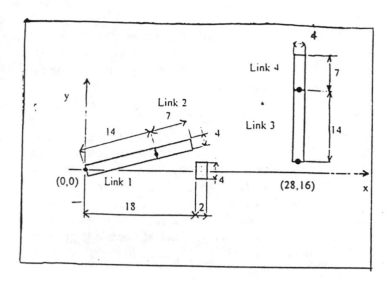

fig.8
Geometric data for two link arms example

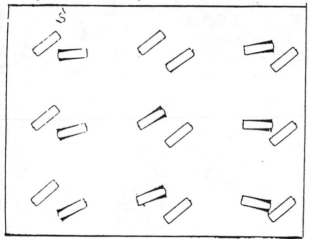

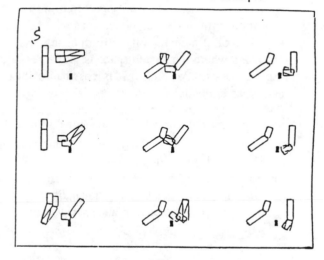

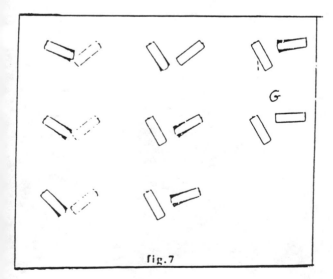

fig.7

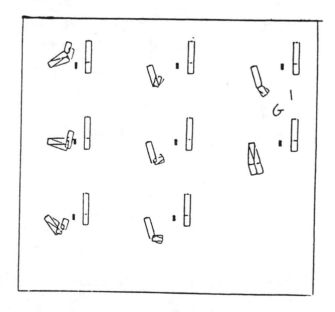

fig.9

47

5. CONCLUSION

A CAD procedure for collision-free path-planning has been proposed and analysed. Its key ingredients are the computation of the Cspace and the construction of a suitable distance-field. The possibility of employing such procedure in the case of multiple manipulators acting in the same environment with obstacles has been tested and good performances have been achieved.

The major difficulty with an approach of the above kind is in the dimensionality of the Cspace. The dimension of the Cspace for two manipulators having respectively n_1 and n_2 degrees of freedom is given by

$$N = \prod_{i=1}^{n_1+n_2} \frac{\Delta\theta_i}{d\theta_i}$$

, where $\Delta\theta_i$ is the range of variation of the i-th joint variable and $d\theta_i$ is the corresponding discretization step. For two six degrees of freedom manipulators, with $\Delta\theta_i = 180$ and $d\theta_i = 18$, we obtain $N = 10^{12}$, a dimension clearly prohibitive. We give a condition to limit Cspace dimensions with (1), but the problem of dimensionality still remains. As in all problems where the search space is too large we must use heuristics. We think that is possible to introduce heuristics in the developed method, simply considering a variable discretization step. The idea is quite simple: we begin by choosing big values for $d\theta_i$ and plan a path; then we successively refine the found path using lower values for $d\theta_i$. In this way we need not to memorize the entire Cspace. This heuristic seems to give us a good solution of the problem, in that we can choose between completeness and execution time.

REFERENCES

[1] **B. Ravani**
"CAD based programming for sensory robots"
NATO ASI series F50, Springer Verlag, Berlin, 1988.

[2] **M. Sharir**
"Algorithmic motion planning in robotics"
Computer IEEE, March 1989.

[3] **T. Lozano-Perez**
"Spatial planning: a configuration space approach"
IEEE Transactions on Computers, Vol.C-32, 1983, pp. 108-120.

[4] **Y. K. Hwang,**
"Boundary equations of configuration obstacles for manipulators"
Proceedings of IEEE International Conference on Robotics and Automation, pp. 298-303, 1990.

[5] **P. Adolphs, D. Nafziger,**
"A method for fast computation of collision-free robot movements in configuration-space"
IEEE International Workshop on Intelligent Robots and Systems, pp. 5-12, 1990.

[6] **A. Rosenfeld, A.C.Kak,**
"Digital picture processing"
Academic Press,1976.

A CAD—BASED ROBOT PROGRAM GENERATOR FOR A WELDING APPLICATION

Michael Cargnelli* and Adam Rogowski**

*Austrian Research Centre Seibersdorf GmbH, A-2444 Seibersdorf Austria
**Technical University Warsaw, Poland

Abstract: A computer program to generate arcwelding robot program code from an input of part geometry and weld-description CAD-data and additional rules and parameters was developed for an Austrian manufacturer structural steel work. The system is needed because a large variety of products has to be manufactured in small batches, so that on-line programming (teach-in) becomes a serious cost factor and limits the throughput.

Key Words: Arcwelding; off-line programming; automatic program generation

1. INTRODUCTION

Typically, robots are programmed using a teach-pendant to guide the robot through a sequence of positions and operations. Although interactive and easy to learn, this method requires the use of a robot as a programming tool, tying up a valuable resource. In case a limited variety of products is run for months or years, as is usual for example in automobile production, the programming time is still negligible compared to production time.

On the other hand, if products are designed for a specific customer and manufactured in small batches, the running time with one robot program may be as short as a few shifts. If this is the case, and the application is not extremely simple, on-line programming time may come up to the same order of magnitude as production time. Under such circumstances an investment in industrial robot systems will never pay off.

To open these new fields of robot application it is unavoidable to face the problem of automatic program generation, since we believe

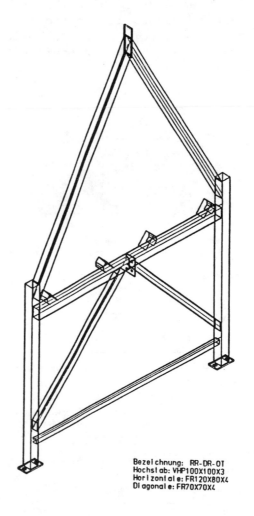

Bezeichnung: RR-DR-01
Hochstab: VHP100X100X3
Horizontale: FR120X80X4
Diagonale: FR70X70X4

Fig.1: Example of steel structural

49

that, in general, graphical teach-in is not the way out.

So the objective is: "Find a way to formalize the skill, the experience and the implicit applied knowledge of the human robot programmer, to enable a computer program to create applicable robot programs."

2. THE SYSTEM

The CAD-data is provided by SUN-workstations running an application for steel construction on basis of the CADDS system and then transferred to a VAX/VMS computer via LAN. Our application is running on this machine. The resulting robot programs are then transferred to a PC right beside the welding facility. There ABB's Off-Line Programming Package is used to compile the program and download it to the controller. The robot is an ABB IRB-2000 with ESAB welding equipment and SMARTAC sensor.

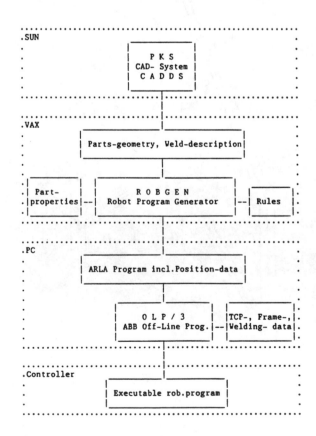

Fig.2: System integration

The following tasks have to be performed by the program generator:

1. Read part geometry and weld description. Associate parts with part property data held in a database.

2. Resolve ambiguities in weld description by applying rules from an interactively changeable rule-table to define which sides of the connections are to be welded.

3. Calculate weld-seam coordinates and tool-orientations using the results from (2) plus various parameters.

4. Calculate search paths for utilizing the adaptivity function of the robot.

5. Get arcwelding parameters by applying rules from a 2nd rule-table.

6. Calculate the robots external axes positions by comprising weld-seams which are accessible from the same robot base position.

7. 2-D display of parts, tool-orientations and search pathes. The user may now modify some data interactively, causing recalculation of the changed seams using the altered parameters.

8. Generate ARLA robot program code.

There is no need to tackle the problem of collision free paths in full universality since the structure is quasi 2-dimensional. Nevertheless some parts require rather intricate strategies to find a feasible tool orientation.

The rule processor uses rule tables, which can be changed interactively. Each rule table element contains "OR" conditions concerning one property of the connection e.g. the allowed part-types for part-1. A rule (a row in the table) has fired when the logical "AND" of all set (nonblank) column elements holds.

```
POS V=60% X=540.0 Y=-760.0 Z=1706.0 ->
    ZZ=-120.0 YY=-22.5 XX=0.0 ->
    E1=843219 E2=-82115
CALL PROG 300
POS V=33% FINEC X=540.0 Y=-760.0 Z=1910.0 ->
    ZZ=-120.0 YY=-22.5 XX=0.0 ->
    E1=843219 E2=-82115
MODRECT COORD
CALL PROG 301
POS V=1% AUTOSEARCH S11 ->
    X=568.0 Y=-760.0 Z=1910.0 ->
    ZZ=-120.0 YY=-22.5 XX=0.0 ->
    E1=843219 E2=-82115
CALL PROG 302
ROBOT COORD
POS V=30% X=538.0 Y=-760.0 Z=1904.0 ->
    ZZ=-120.0 YY=-22.5 XX=0.0 ->
    E1=843219 E2=-82115
CALL PROG 303
POS V=60% X=538.0 Y=-760.0 Z=1628.0 ->
    ZZ=-120.0 YY=-22.5 XX=0.0 ->
    E1=843219 E2=-82115
MODRECT COORD
POS V=100% X=580.0 Y=-222.6 Z=1736.5 ->
    ZZ=-60.0 YY=-82.0 XX=160.0 ->
    E1=843219 E2=-82115
POS V=60% FINEC AWELD 1/13/4/0 ->
    X=580.0 Y=-780.0 Z=1900.0 ->
    ZZ=-60.0 YY=-82.0 XX=160.0 ->
    E1=843219 E2=-82115
POS V=10% FINEC WEND 1 X=580.0 Y=-780.0 Z=1980.0 ->
    ZZ=-60.0 YY=-82.0 XX=160.0 ->
    E1=843219 E2=-82115
```

Fig.3: Sample of generated ARLA program

3. CONCLUSIONS

The system is currently in the test phase at the customer's plant. First experiences from real application are promising. So it seems, that indeed the mostly automatic generation of robot program code might open new fields of robot application in situations, where up to now industrial robot equipment could not be used, because of low production efficiencies resulting from on-line programming.

4. REFERENCES

M.Naval: Roboter-Praxis; Würzburg: Vogel Buchverlag, 1989

CAD Based Programming for Sensory Robots; Ed. B.Ravani. NATO ASI Series F, Vol.50, 1988

COMPUTER AIDED PLANNING OF FLEXIBLE, MODULAR ASSEMBLY CELLS

D. Noe and P. Kopacek

Department of Systems Engineering and Automation
Scientific Academy of Lower Austria
Krems, AUSTRIA

Abstract: Assembly automation in form of robotized assembly cells is one of the most important application areas of robotics today and in the future. The flexibility of these cells offers possibilities to automatize assembling of small and medium lots. Planning of assembly cells will be done manually in most cases by means of some computer facilities. The paper presents first results for totally computer aided planning. The creation of an integrated database model, the planning process and knowledge will be discussed. The assembly operations, their sequencing, list of selecting components, assembly time and price of flexible cells are the goals of the proposed planning system.

Keywords: Robotics applications, Assembly, Assembly Automation, Assembly cells.

1. Introduction

In this paper some ideas and experiences are presented dealing with "intelligent" automation especially in small and medium sized companies. Such companies are very important for the productivity in various mainly smaller coutries. For example in Austria as well as in Slovenia aproximately 50% of the industrial production is carried out in companies with less than 500 employees. Compared with larger companies one of the main advantages is flexibility. Small companies are more inclined to produce special products according to the demand of the market in small lots. Doing this in an efficent way, flexible automation is absolutely neccessary today and in the future. Computer integrated manufacturing (CIM), and its elements (CAD, CAPP, CAM etc.) support the flexibility and are also one of the solutions for small and medium sized companies with small lots production.

Growing efforts have been observed in the last decade to systematize the assembly system planning process and to support the planner at planning decisions. Some important steps were taken in planning process systematization (Bullinger 1989, Scholz 1990), in assembly structure planning (Ammer 1985), in assembly operation optimisation (Eversheim 1989, Schöninger 1989) and expert systems involving in process planning (Eversheim 1989, Feldmann 1990).

The planning and the development of a flexible, robotized assembly cell is carried out mostly manually today. According to an increasing demand the development of computer aided systems to support the planner in building optimal configured cells at lowest time consumption is necessary.

The assembly operation and operations sequencing determination, the selection of robots and the cell modules, the calculation of the assembly time and the animation and simulation of the assembly process are typical tasks to be performed in setting up a computer aided planning system for robotized assembly cells. Such a CAP system consists of a large relationed database containing part description, cell modules as well as the software modules for part selection, for layout design, time determination and simulation. The system offers the possibility of optimizing the arrangement of peripherical devices for minimizing the assembly time and the assembly cell cost. A kind of expert systems for robot and peripherical devices selection supports the decisions for an efficient development of the assembly cells.

This work was supported by the Austrian "Forschungsförderungsfonds für die gewerbliche Wirtschaft - FFF."

2. THE PLANNING SYSTEM

The development of the planning system is based on the assumption that flexible assembly cells consist of a group of modules, such as robots, grippers, delivery and orienting devices, various assembly tools, sensors and control, to work together simultaneously on certain assembly operations. An assembly cell consists at minimum of one robot equipped with one tool or gripper. It should perform a sub assembly or product assembly of one or various products in flexible assembly cell. Several cells are connected to the the assembly lines or assembly systems trough a transport system.

Different effective planning methods are known in the assembly systems planning and the planners disposal of various support tools (Bullinger 1986). In the development of the planning system those methods were briefly surveyed and taken into consideration.

In the development of the planning system the following tasks are taken into account:

• creation of a database and database management system,
• determination of the planning system goals,
• the planning flaw or planning system building,
• collection of neccessary data and knowledge,
• testing of the planning system.

3. INTEGRATED DATABASE

The main part of a successful computer aided system for planning flexible assembly cells is an integrated database. From the complexity of the assembly planning process follows a special developped database. Various data were collected for planning reasons and stored into the database as shown in Fig. 1.

• the information about assembled products, parts and assembly process, the *customer database*,
• the information about assembly cells modules, *component database*,
• the information about planning results, *assembly planning process database*,
• the information and knowledge to support the planning decisions, *knowledge database*.

The *customer database* contains informations to describe the assembly cell tasks as well as the demands for useful applications. The product structure, the part description such as the geometrical and technological part properties, number of parts of a product, the parts and product dimensions, the way of gripping, part delivery state, part and product quality data, product variants, the assembly cell integration in the whole proguction as well as the integration in the production control system and the assembly cell performance are some of technical customer informations. The customer database contains also economic informations, such as the assembly cell price, the date of delivery and also the requested

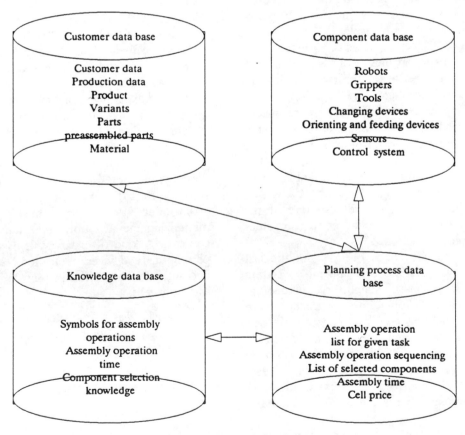

Fig. 1: Integrated database

assembly cell time. The customer database is created always for each application and the data are in the relation to the application.

In spite of a great number of different assembly devices which are necessary for building an assembly cell, it is possible to systematize the modules and create the appropriate *component database*. In comparison with other computer

assembly planning systems (Eversheim 1989, Feldmann 1987) the planning process takes into account different modules available on the market as well as the self-produced modules. The relevant informations needed for component selection are stored in the relation of component database. The information consists of the general data, technical data and commercial data. The selected component has to satisfy the technical as well as

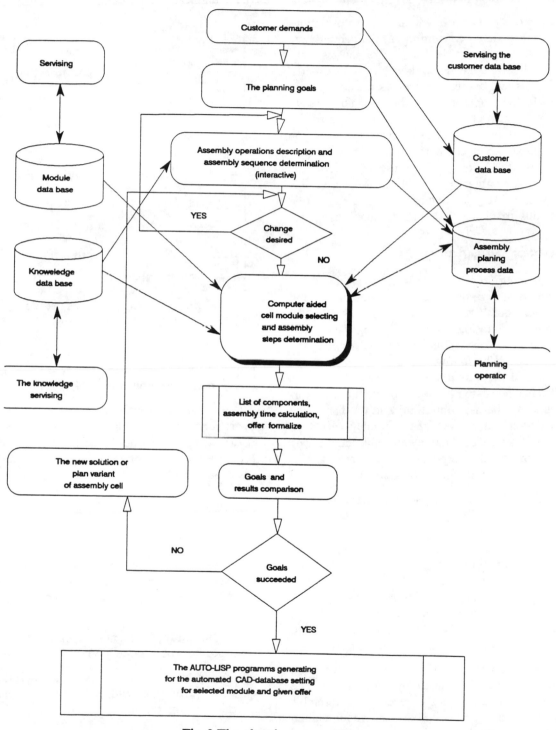

Fig. 2: The planning process flow

the economical demands.

Planning process database is related to the application problem. The results of each planning step are stored to be used as information in the later planning tasks. The informations items are the assembly operations, assembly operation sequencing, assembly time, selected components as well as the layout for assembly cells, graphical representation and movement simulation.

The *knowledge database* contains the informations about the time consumption performing the assembly operation. The informations are collected from the assembly planning experts, catalogues and manuals as well as calculated from the selected components. The planning system reduces the components selection to the comparison of the assembly operations and the components which can perform the given assembly operations. The final choice must be made by the planner. A symbolic presentation of the assembly is introduced for planning reasons. The calculation equations are also included in the knowledge base.

4. THE PLANNING PROCESS

The planning process starts with the products and their part description, the goals definition, and customer data base creation (Fig. 2). The assembly operations determination and their sequencing is done manually. As the result of the processing, in the following planning step, a rough proposal of the assembly cell is made. It consists of the needed assembly operations, the assembly sequencing, the assembly time and the list of the chosen component with the assembly cell price and delivery time. The layout and the moving simulation of the proposed cell are the task of following steps in the planning system.

The assembly operations are described by standardized assembly operations {$AO1,....AOi$}, such as: part inserting, part screwing, product manipulating, testing the assembled product. The standardized assembly operations for a robotised assembly cell are stored in the knowledge data base. Each assembly operation is represented by a symbol and the assembly minimum and maximum time is given as well as the component group which can perform the assembly operations.

The assembly process and the assembly operation sequencing are determined by the planner. The planner chooses the necessary assembly operations, the part number, the manipulation operations. Then the part is automatically selected from the part database, gives the number of the assembly operation repetition, and the needed number of robots.

The assembly operations and the assembly operation sequencing are graphically presented by symbols (Fig. 3). The chosen assembly operations are necessary to assemble the whole product. The choice is based on the assembly operations {AOi} and demands two robots working symultaneously. The distribution of the assembly operations and the operation sequencing have been carried out by the planner.

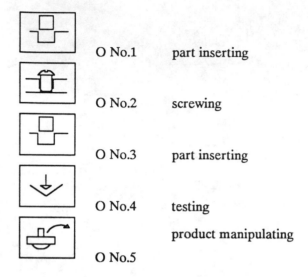

	O No.1	part inserting
	O No.2	screwing
	O No.3	part inserting
	O No.4	testing
	O No.5	product manipulating

Fig. 3: The assembly process presentation

The assembly operations, performed by robot, are executed by one or more grippers {$G1,....,Gi$} and one or more assembly tools {$T1,.....Ti$}.

The selection of grippers is based on the assembly operations, part gripping and dimension demand. If the assembly operation can be executed by two gripper types, the planner, according to the all assembly operations, has the chance to choose only one type, for example the vacuum gripper. For a given part dimension a group of grippers is available. The final selection is done by the planner with the knowledge:

- if there are two or more assembly operations executed by the same gripper type and the same finger type, choose the gripper which can grip both part.

- if a maximum of assembly operations should be executed by the same gipper,

- for the two sequenced assembly operation the same gripper should be used to avoid the changing operation.

The planner's task is to choose minimum number of gripers and assembly tools considering the lowest price and the shortest assembly time. The planning system supports the planner with stored data and graphical presentation.

The standardised assembly operations are composed by different assembly steps {AOS1,...AOSi}. The assembly operation can be carried out with some of the following steps : tools or grippers changing, gripper setting, tools or grippers feeding, part preparation, part feeding and assembly operations execution.The selection of the robot and part preparation equipment and time calculation are based on the determination of assembly operations steps.

Robot selection parameters are: number of the axes, the load, controlling,accuracy.

Corresponding to those selection parameters, some robots are chosen {R1,....,Ri}, and the planner has the opportunity to make the final decisions.

There are some different feeding and orienting equipments available {FOE1,....FOEi}. The selecting is based on the following relevant parameters:

- the delivery conditions of the parts and subassembly products,
- part geometrical properties (the part shape group),
- part dimensions,
- part sensitiveness,
- the assembled part per hour.

The planner has the oppurtunity to make the final decision. The selection is also supported by the calculations, the delivery part numbers and the calculation of the manual part handling time.

For assembly time calculation the demands are the assembly operations steps time calculated and stored in the planning database. The assembly operations data are collected from manuals, catalogues and planner's expiriences and some times they are also measured. The feeding time is calculated related to the robot moving speed and the path length. Since the lay-out is not present ay the moment, the path length is defined by the planner (Kopacek 1992).

5. SUMMARY

Assembly automation is absolutely necessary especially for small and medium sized companies. Therefore robotized assembly cells should be introduced more and more in production processes. Such cells ensure a high flexibility necessary for the factory of the future which the the just in time production will predominate.

The planning and the development of such assembly cells is a time consuming process today. Therefore a computer aided planning system as an efficent tool for reducing the planning time was developped. The system supports the planner, and shortens the planning time remarkable. Main components of the planning system are the component, customer and knowledge database and the interactive planning system.

The important planning system's attributes are:

- the flexible assembly cell task is descibed with the standard assembly operation and the assembly
- operations steps,
- the component selection is based on assembly operations and assembled parts properties,
- the assembly time optimization is possible,
- the concept of the planning system allowed the extension of the knowledge base in order that the system can take over more planner's tasks.

This computer aided "intelligent" planning system for robotised assembly cells will be implemented in this year in an Austrian small-sized company.

6. References:

E.D. Ammer: *Rechnerunterstützte Planung von Montageablaufstrukturen für der Sereienfertigung*, Springer Verlag, Berlin 1985

H.J. Bullinger: *Systematische Montageplanung*, Carl Hanser Verlag Münhen 1986

W. Eversheim: *Expertsysteme und Simulation zur Planung von Montagesystemen*, Technische Rundschau 23/1989, S 24 - 29

W. Eversheim; H. Esser; F. Lehmann: *Computerunterstützte Montageplanung und Steuerung, Durchlaufzeit reduzieren*, Industrie-Anzeiger 68/1990 S. 14 - 15

K. Feldmann, A. Hemberger: *Rechnereinsatz in der Montageplanung*, VDI- Z. 129 (1987), 5, S. 76 - 81

W. Scholz: *Einsatz von Datenbanken bei der rechnergestützten Planung von Montageanlagen*, CIM Management 1990, 4, S. 18 - 23

J.Schöninger: *Planung taktzeitoptimierter flexibler Montagestation*, Springer Verlag, Berlin 1989

P. Kopacek, K. Fronius: *CIM Concept for the Production of Welding Transformers*, In Preprints of "INCOM '89, Vol.2,p.737-740

P. Kopacek: *CIM for Small and Medium Sized Companies*, Research report, Austrian Ministery for Science and Technology, Viena 1989 (in German)

P. Kopacek, D. Noe: *CAP for assembling*. Will be published in the Proceedings of the Workshop "Robotics in Alpe-Adria Region", Portoroz, June 1992.

COMPUTER AIDED DESIGN OF NC MACHINE TOOL
MAIN SPINDLE DRIVES

Vladimir DUKOVSKI and Zoran PANDILOV
Department of Mechanical Engineering
The "Cyril and Methodius" University Skopje
Karpos II b.b., 91000 Skopje, MACEDONIA

Keywords: main spindle drives, NC machine tools

Abstract: Charactristics of main spindle drives highly depend upon skilfulness of composing motors and mechanical transmission elements. This paper gives short description of an original computer program which enables interactive design of main spindle drives for NC machine tools and analysis of different design variants.

INTRODUCTION

Contemporary development in machine tools is connected with improvements in drive systems.

A special characteristic of NC machine tool drives is application of variable speed motors which provide continous changing of cutting speeds an feed rates.

Application of variable speed motors creates a question of their appropriate composing with mechanical transmission elements in order to get better output characteristics of the main spindle.

CHARACTERISTICS OF THE MAIN SPINDLE DRIVES FOR NC MACHINE TOOLS

Main spindle drives for NC macnine tools must provide constant power at wide range of speeds on the output of the main spindle.

They consist of three parts:
1. variable speed motor,
2. mechanical transmission elements which provide appropriate output characteristics of the main spindle,
3. main spindle.

Usually mechanical transmission elements consist of:
- belt transmission,
- combination of belt transmission with gearbox (with two, three or four speeds).

Intensive development of quality tool materials enable using of very high cutting speeds and power.

Necessary output power on the main spindle can be calculated as:

$$P = \frac{F_t \ v}{60 \ 10^3} \ [KW], \quad (1)$$

F_t- tangential cutting force component [N],
v - cutting speed [m/min].

NC machine tools are used for production of parts with different shapes, dimensions and materials, with wide range of cutting conditions.

For ensuring these requirements the speeds on the main spindle must be regulated in very wide range,

$$R_{ms} = \frac{n_{max}}{n_{min}} = \frac{V_{max}}{V_{min}} \frac{D_{max}}{D_{min}} = R_v R_d, \qquad (2)$$

R_{ms}-range of regulation of output main spindle speeds,
R_v -range of regulation of cutting speeds,
R_d -range of diametars of the parts or cutting tools,
n_{min}, n_{max} -maximal and minimal main spindle speed,
V_{max}, V_{min} -maximal and minimal cutting speed,
D_{max}, D_{min} -maximal and minimal diametars of the parts or
        cutting tools.

According to our empirical investigation the range of regulation of main spindle speeds for NC machine tools, usually is within R_{ms}=20-350 (exclusively rare to 600). Such kind of wide regulation of main spindle speeds needs particular attention in selection of variable speed motors and mechanical transmission elements.

The reasons for using these variable AC and DC motors are that they offer a higher degree of automation, high workpiece quality, shorter machining and idle times during the machining process, longer tool life and low operating noise.

Fig.1 presents power-speed diagram of variable speed motor, where n_{mmin}, n_{mn} and n_{mmax} are minimal, nominal and maximal speed of the motor, and P_m is nominal power of the variable speed motor.

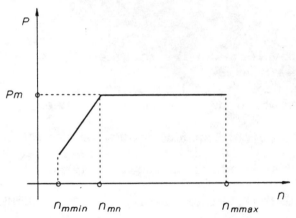

Fig.1: Power-speed diagram of variable speed AC motor

Usually the range of regulation of speed at constantpower of variable speed motors is (2-8) (sometimes reach values 12-16) which is far bellow required range of R_{ms}=20-350.
On the other hand

$$R_{ms}=R_{msm} R_{msp} , \qquad (3)$$

R_{msm}=2-50 -range of regulation of output main spindle speeds at
        constant torque,
R_{msp}=2-45 (exclusively rare 70)-range of regulation of main spindle
        speeds at constant power.

There are two alternative methods of obtaining wide range of main spindle speeds at constant power:
        -overrating of the AC or DC motor,
        -combining the motor with gearbox with two, three or four speeds.

Overrating the variable speed motor is not a cost-efective solution because:
        -the drive installed power is not fully utilized,
        -the operating costs are increased,
        -the motor is physically large,
        -the power supply system must be made stronger that is necessary.

Overdimensioning of the AC or DC motor is an economic solution up to approximately 10 [KW] /6/.

The second solution with two, three or four speed gearbox is widely used at the NC machine tools.

Selecting the number of steps Z of the gearbox is based on the range of regulation of the variable speed motor with constant power R_{mp}, while with using the range of variable speed motor with constant torque $R_{mm} = R_{msm}$, the whole range of regulation of output speeds of the main spindle is obtained.

Because of that, we can write:

$$R_{msp} = R_{mp} \, R_z \quad (4)$$

R_{mp} —range of regulation of variable speed motor with constant power,

R_z —range of regulation of the gearbox.

Variable speed motor can be treated as a particular group of gearbox with continuos changing speeds, which is first in the kinematic chain, with infinitely large number of transmissions, with transmission ratios which obtain geometrical progression with progression ratio $\varphi \rightarrow 1$ and range R_{mp} /4/.

Gearbox can be treated as a transmission group which extend the speed range of the motor at constant power. Because of that characteristic of a transmission group φ is:

$$\varphi_z = R_{mp} \, \varphi \quad (5).$$

Because $\varphi \rightarrow 1$, we obtain

$$\varphi_z = R_{mp} \quad (6).$$

As,

$$R_z = \varphi_z^{(z-1)} \quad (7),$$

we can write

$$R_z = R_{mp} \quad (8).$$

If we replace (8) in (4), we get

$$R_{msp} = R_{mp} \, R_{mp}^{(z-1)} = R_{mp}^z \quad (9),$$

z-number of speeds of the gearbox.

When R_{msp} and R_{mp} is known, we can easy estimate the necessary number of speeds of the gearbox, using the equatation (9). Then we get:

$$Z_0 = \frac{\log R_{msp}}{\log R_{mp}} \quad (10),$$

Z_0- calculated number of speeds of the gerbox.

The equatation (10) is often recommended in the literature /2,3,4/, for calculation of number of gearbox speeds.

Because Z_0 is usually a decimal number, it is rounded to the nearest integer number (Z).

If $Z > Z_0$ we get structure with overlapping the speeds.

In case of $Z < Z_0$ we get structure with local definite decrease of the power δP.

For example if the result from (10) is $Z_0 = 2.5$, than Z can be 2 or 3. In case of Z=3 obtained P-n daiagram is as on the fig 2a), and if Z=2 obtained diagram is as on the fig. 2b).

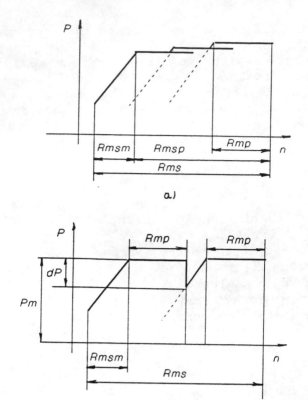

Fig.2: Diagram P-n of the main spindle
a) with z=3 and b) with z=2

Procentual decrease of the power dP in relation with the nominal power P_m of the motor, when $Z < Z_o$, can be calculated with the equatation (11) /5/,

$$\frac{dP}{P_m} = \left[1 - R_{mp} \sqrt[z-1]{R_{mp}/R_{msp}}\right] 100 \, [\%] \qquad (11).$$

Usually dP/P_m should not be greater than 30% /5/. Similar results were obtained in our empirical investigation.

DESCRIPTION OF THE COMPUTER PROGRAM FOR DESIGNING MAIN SPINDLE DRIVES

Theoretical considerations, mentioned in the previous chapter are implemented in the computer program. An original computer program for interactive design of main spindle drives and analysis of different design variants was created for PC in C-language.

Flow chart of the computer program is given on fig.3.

The program begins with input of tangential cutting force component Ft and cutting speed v. They are necessary for calculation of required power. Existes a posibility to enter directly required power for particular size of NC machine tools, based on recommendations implemented in the computer program. Recommendations are result of of the empirical investigations of main spindle drives of NC machine tools. More then 2000 different NC machine tools were investigated and appropriate recommendations were derived.

In the next step computer program selects variable speed motor from the AC/DC motor database. For the selected motor the program draw power-speed (P-n) and torque-speed (M-n) diagrams. The P-n and M-n diagrams are shown on fig.4.

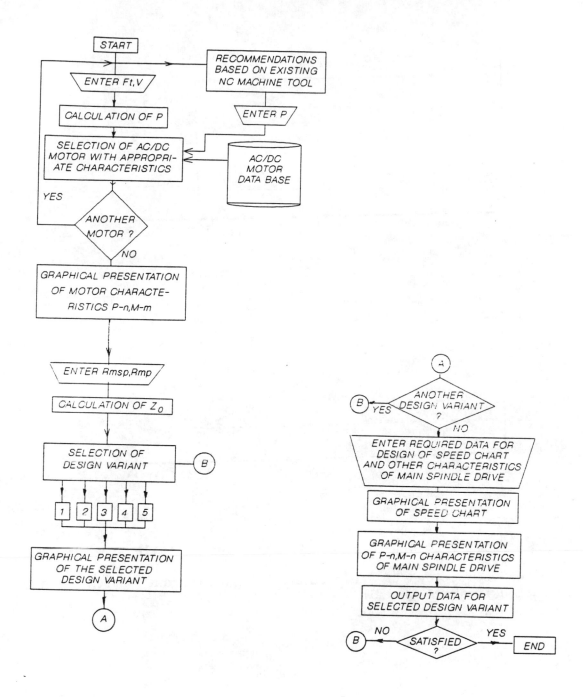

Fig.3: Flow-chart of the computer program

In the next step the values of Rmsp-range of regulation of main spindle speeds at constant power and Rmp-range of regulation of the variable speed motor with constant power are required. This is necessary for calculation of number of speeds of the gearbox Zo.

Then the computer program gives us oportunity to select one of the most frequently used design variants of main spindle drives for NC machine tools:

-1. motor-belt-main spindle,
-2. motor-planetary gearbox-belt-main spindle,
-3. motor-belt-gearbox-main spindle,
-4. motor-belt-gerarbox-belt-main spindle,
-5. motor-belt-reducer-main spindle.

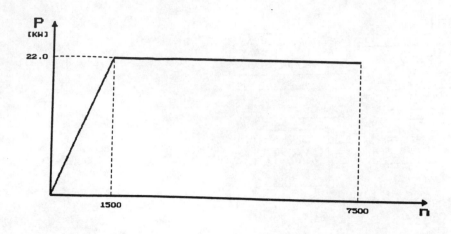

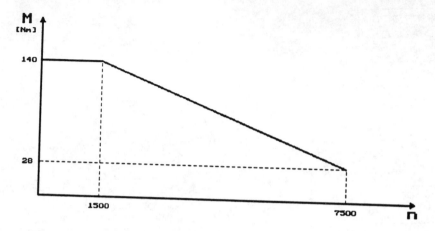

Fig.4: P-n and M-n diagram for the selected variable speed motor

After selection of particular design variant its graphical presentation is shown (fig.5).

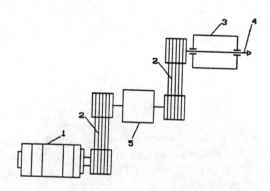

Fig.5: Graphical presentation of design variant
motor-belt-gearbox-belt-main spindle
1. variable speed motor, 2. belt transmission, 3. spindle unit,
4. main spindle, 5. gearbox (z=2,3 or 4)

In the next step elements for drawing speed chart (transmission ratios in the gearbox, transmission ratio(s) of the belt transmission(s) etc.) are required.

Follows graphical presentation of speed chart (fig.6), power-speed (P-n) and torque-speed (M-n) diagrams (fig.7a and 7b) and textual presentation of the output data for selected design variant (fig.8).

If we are not satisfied with the output data we can select another design variant.

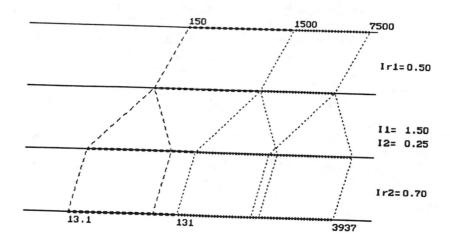

Fig.6 Speed chart for the selected design variant
motor-belt-gearbox (z=2)-belt-main spindle

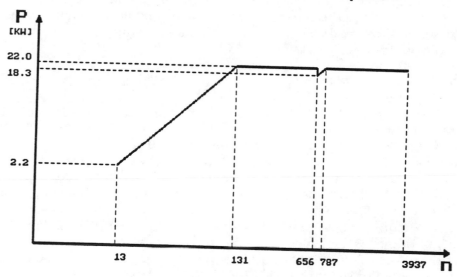

Fig.7a: P-n diagram of the main spindle

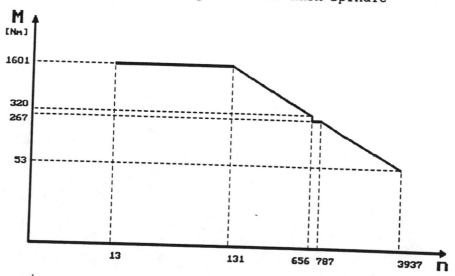

Fig.7b: M-n diagram of the main spindle

OUTPUT DATA FOR SELECTED DESIGN VARIANT
motor-belt-gearbox (z=2)-belt-main spindle

```
Nmin-minimal main spindle speed        Nmin=13.1   [1/min]
Nmax-maximal main spindle speed        Nmin=3937   [1/min]
P   -maximal main spindle power         P=22       [KW]
Pmin-minimal main spindle power        Pmin=2      [KW]
Mmax-maximal main spindle torque       Mmax=1601   [Nm]
Mmin-minimal main spindle torque       Mmin=53     [Nm]
Rms -range of regulation of the main spindle
                                       Rms =300.00
Rmsp-range of regulation of the main spindle at constant power
                                       Rmsp=30.00
Rmsm-range of regulation of the main spindle at constant torque
                                       Rmsm=10.00
```

Fig.8: Output data for the main spindle design variant

CONCLUSION

Characteristics of main spindle drives for NC machine tools directly depend upon skilfulness of composing variable speed motors and mechanical transmision elements. The presented computer program enables interactive design and analysis of different variants and reduces the time for design of main spindle drives.

REFERENCES

[1] Z. Pandilov, V. Dukovski, Lj. Dudeski: Design of main spindle drives for NC machine tools,Proceedings of the Faculty of Mechanical Engineering-Skopje, No.10, pp.159-170, 1991.

[2] V. Dukovski, Lj. Dudeski: Machine tool design, Skopje,1989.

[3] V.E. Push: Machine tools, Mashinostroenie, Moscow,1986.

[4] N. Acherkan: Machine tool design, Mir Publishers, Moscow, 1982.

[5] B.F. Epifanov, B.M. Krilov: Selection of motors and optimal parametars of gearbox for lathes, Stanki i instrument, 1978, No.2, pp.13-14.
[6] Winkelmann S., Knur W.: Extending the Speed Range of DC Main Spindle Drives for Machine Tools by Means of Mechanical Gearboxes. Siemens Power Engineering III(1981) No.11-12, pp.318-321.

[7] Knur W., Winkelmann S.: A Two-Speed Gearbox and a DC Main Spindle Motor in a Compact Unit. Siemens Power Engineering PRODUCT NEWS 3 (1983) No.2, pp.50-51.

[8] L. Ammeraal: Graphics Progrming in Turbo C. John Wiley & Sons, Chicester, 1990.

[9] H. Schildt: Turbo C, The Complete Reference, Mc Graw-Hill, 1987.

SOLID MODELLING FOR NC MACHINING ANIMATION

Vesna ARNAUTOVSKI[1], Dragan MIHAJLOV[1] and Vladimir DUKOVSKI[2]

[1]Faculty of electrical eng. [2]Faculty of mechanical eng.
The "Kiril i Metodij" University Skopje
Karpos II bb 91000 Skopje, Republic of Macedonia

Keywords: solid modelling, NC machining verification

Abstract: Inevitable part of CIM environment is modelling, simulation and animation of objects, processes and systems. Computer animation of the NC machining represents the actual tendencies in visual verification of NC programs. In the paper a concept of NC milling PC animation is presented by introducing its principal ideas and application aspects. It is developed as an inexpensive PC test model for graphical simulation of the machine tool cutting action. Special attention is given to dynamical display of the tool-workpart system during the machining, thus verifying the metal removal process of an NC tool path.

INTRODUCTION

In machining processes, for example cutting operations, NC machine tools represent a well established technology. But still largely high cost of controls and NC programming exist. One of the approaches to drop significantly the costs is to approve the process of NC programming more efficient and easier. There are languages for specification of tool paths which often include graphics simulation capability for tool path verification. Another method of generating NC programs is available on some computer aided draughting systems which allows the designer to input a tool path over a drawing of the object. The advantage of being able to generate NC programs directly from geometric models has been recognized for some time, though the results so far are limited. Another level of application in milling operations is the verification of an existing NC program (Encarnacao 1980). There are already a lot of graphics simulations available which are based on geometric modelling and are used:
- to verify that only surface contact takes place between the swept out tool volume in its path and the model;
- to subtract the swept out tool volume from the stock material and to check if the remaining object is indeed the desired object.

In the paper a concept of NC milling PC animation that uses an existing NC program to produce graphical output for its verification is presented. Special attention is given to solid modelling for an object generation and animation of the machining process. The paper is organized as follows. In the first section global characteristics of the system are presented. In the following sections special attention is given to solid modelling and animation. On the end some software characteristics are given.conclusion is presented.

GLOBAL CHARACTERISTICS

As a result of collaboration between faculties of Mechanical and Electrical engineering a system for graphical verification of machine tool cutting action is introduced. Its objective is to provide easy and inexpensive way to test NC programs without using a machine tool. The NC tool path data are used for generation of dinamical image of the machining process. The system is structured into two functionally different parts (fig.1). The first part includes functions for NC program analysis and toolpath geometry calculation. The second part generates animated solid model of the workpiece and tool during machining and graphical output of the machining process.

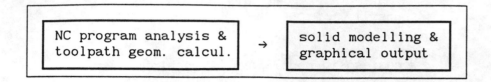

Fig.1: Global structure of the system

NC program analysis & toolpath calculation

The NC program analysis is the first step for graphical interpretation of the machining process. Data containing geometrical information about the stock material, tools and tool path are selected. Also, information about tool compensation is taken into account. Finally, the tool path geometry is calculated and information is saved as internal data file (fig.2).

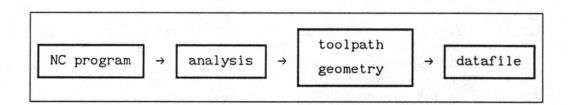

Fig.2: Global structure of the NC program analysis

Solid modelling & graphical output

Visual interpretation of the NC program is based on the internal data file. Data is used for calculation of the swept out tool volumes and generation of CSG tree.

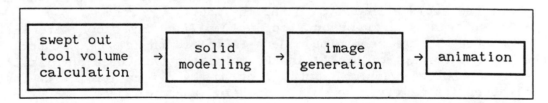

Fig.3: Global structure for graphical output generation

In the next step the screen domain coordinates transformation is performed and B-rep presentation is generated. The modifications in the CSG tree are transferred only in local screen area (Bronsvort 1989).On the base of the previous image, the new image is generated introducing corrections only in the area of interest as a result of modifications in visibility of the edges and surfaces of the new model. On the base of these data an animation cycle is generated. The global structure of this process is presented in fig.3.

In the following sections special attention is given to solid modelling process and generation of an animation cycle.

SOLID MODELLING

The workpart machining is represented using geometric modelling for incrementally modifying the initial model of the object. It means that the stock material is modeled using the volume swept out by the cutting tool as it follows the motion specified in a NC part program. This volume is subtracted using Bolean operations from the stock volume. On this way we have a model of the object in process during various stages of machining. The solid modelling concept is based on hybrid CSG modelling using B-rep presentation of primitives (Rooney 1987),(Encarnao 1980). The modelling process is represented by the CSG tree which indicates how the primitive (swept out tool volumes and the stock material) objects are combined. The characteristics of NC milling process are deduced to binary linear tree (fig.4) containing no explicit information about the shape of the swept out tool volumes, so the resulting shape of the composite object has no explicit information about the vertices, edges and surfaces.

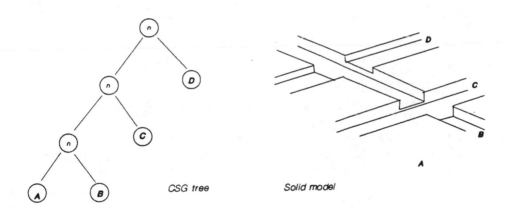

CSG tree   Solid model

Fig.4: Linear binary CSG tree of NC milling process

Boundary representation in screen domain is used to express explicit information about faces, edges and vertices of the primitives. Data are represented in screen coordinates including some information about the tool movement using special structures. Information are calculated using data selected from the NC machining program that is analyzed using preprocessing procedure. Stock material screen coordinates are calculated using the NC program geometry statements (fig.5). For the shape of swept out tool volumes the motion statements are analyzed to select the screen coordinates. The structure contains basic data about the swept out tool volume. For example, a segment of NC program of a straight line milling is analyzed to show the swept out tool volume and the selected data (fig.6).

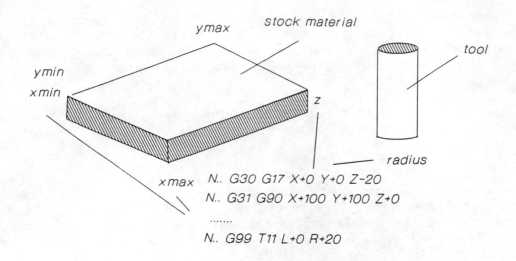

Fig.5: Definition of stock material

The structure contains some screen data including information about the orientation and position of the swept volume with respect to workpiece.

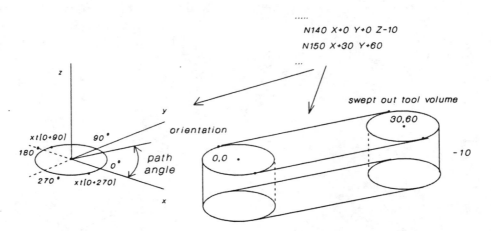

Fig.6: Definition of swept out tool volume

The modelling of work-part is based on the analysis of CSG three after each step of milling using the information selected from NC program. Incremental model modifications during the milling process, represented by the CSG tree, are dynamically displayed using the process of computer animation. It is performed following the basic idea of local image updating. It means that only those screen areas affected by modifications are recomputed and redisplayed.

LOCAL DISPLAY UPDATING

The principle of local display updating follows the CSG modelling as a concept for model visualization. The changes in CSG three as a result of addition of new primitive object are expresses in local screen ares (Broonsvort 1989),(Newman 1973). Only these screen areas are recomputed and redisplayed generating new image of the model. The precise determination of the screen areas, areas of interest,where the changes occur is very important (fig.7). It is based on the analysis of the CSG three using Bolean operators in nodes and B-rep presentation of the primitives (Rooney 1987).

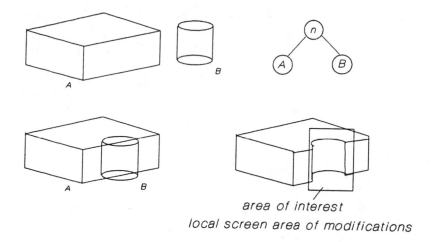

area of interest

local screen area of modifications

Fig.7: Local display updating

The B-rep presentation of the swept out tool volumes is used for intersection points determination between swept tool volume and the work-part. Formally it is deduced to intersection points determination between new and previous swept tool volumes. Several cases are separated (fig.8):
- New toolpath has no intersection with previous toolpath (a)
- New toolpath has intersection with previous toolpath (b)
- New toolpath has no intersection with edges of work-part (c)
- New toolpath has intersection with edges of work-part (d)

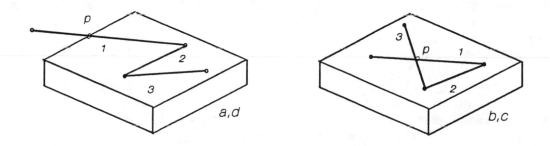

Fig.8: Cases of intersection

A new approach to edge intersection determination and hidden lines removal is introduced using only screen domain data. Its main characteristic is the use of algorithm based on the Bresenham line algorithm for intersection points determination between primitives of CSG and image visibility corrections.

ANIMATION

The NC program verification is presented via 3D solid animation representing the movement of the tool and the metal removal process. The animation is based on the previous and the next image data of the model in the area of interest where modifications in CSG tree are affected. A hybrid technique of XOR bitblt animation and page animation is used (Adams 1987).

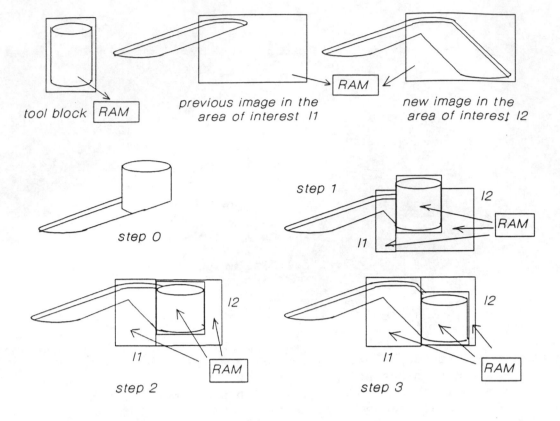

Fig.9: Animation cycle

Bitblt animation uses a pre-defined block to move an image across the screen. A rectangular block of display memory is saved as a RAM array. Different logical operators are used when data are inserted in to the display memory thus creating different image effects. The animation of NC milling is realized using XOR logical operator. The pre-defined block contains the image of the tool which movement is simulated. Also, two pre-defined screen blocks contain the image of the model in the area of interest before and after the modification are used (fig.9). The effect of tool moving across the screen is realized by sequential positioning of the tool block following the toolpath.

SOFTWARE

An experimental software for verification of NC programs is realized on PC using C language. It is used as a prototype that interactively simulates, verifies and displays the metal removal process of an NC toolpath. To begin, a NC program that is to be tested is entered and analyzed. The toolpath data are calculated and the process of solid model generation runs in batch mode. It dynamically shows the motion of the tool and the cutting action. The program can depict both milling and drilling operations with 2.5-axis motion. Tests can be run also one step at a time in interactive mode without displaying the movement of the tool.
On the following fig.10 an example of straight line milling is presented in several steps.

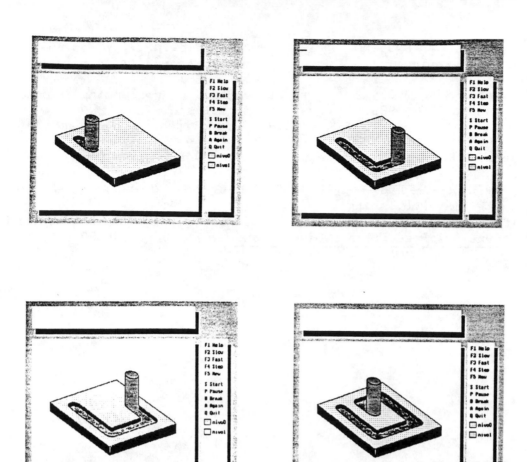

Fig. 10: An example of straight line milling in several steps

CONCLUSION

The main objective of this paper is to introduce inexpensive productive way to test NC programs without using a machine tool. A system for graphical verification of NC programs is presented. Its main function is to produce dinamical visualization of the cutting action in solid modelling technic. It is acceptable for fast program check on-screen by simulating the machining of the workpiece. In this phase it is only experimental program which is still in development.

REFERENCES

Adams L., 1988, "High - performance graphics in C: animation and simulation", Blue Ridge Summit, Windcrest Books.
Alagar V.S., Bui T.D., 1990, "Semantic CSG trees for finite element analysis", Computer-Aided Design", Vol.22, No.4, pp.194-198.
Bronsvoort W.F., Garnaat H., 1989," Incremental display of CSG models using local updating", Computer-Aided Design, Vol.21, No.4, pp. 221-231.
Carbajal J.F., Mendez M.A., 1987, "Edge-edge relationships in geometric modelling", Computer-Aided Design, Vol.19, No.5, pp.237-244.
Encarnacao J. ed.,1980, " Computer Aided Design, Modelling, System

engineering, CAD systems, Springer Verlag, Berlin.

Hagen H., Roller D., 1991, "Geometic Modelling", Sprimger Verlag, Berlin.

Heidenhain TNC 355, 1989, "Operating Manuel, Programming in ISO format".

Heidenhain TNC,,1989, "Programming Examples".

Newman W.M., Sproull R.F., 1973, "Princilpes of interactive computer graphics", New York, McGraw - Hill.

Rogers D.F., 1985, "Procedural elements for computer graphics", New York, McGraw - Hill.

Rooney J., Steadman P., 1987, "Principles of computer - aided design", London, Pitman Publishing.

Wolfe R.N., Wesley M.A., 1987, "Solid modelling for productiondesign", IBM Journal of Research and Development, Vol.3, pp.314-333.

PC BASES ANALYSIS OF CUTTING PARAMETERS INFLUENCE
ON BAND SAW VIBRATIONS

Trposki Z.[1], Mihajlov D., Loskovska S.[2]

[1] Forestry Faculty,
[2] Faculty of Electrical Engeneering,
The "Kiril i Metodij" University Skopje
Karpos II bb 91000 Skopje, Republic of Macedonia

Keywords: primar cutting process, tool vibrations,

Abstract: Quality an quantity of wood primary cutting products are very
important for improvement of semi-final and final wood products. Tool vibrations have a large influence on the quality of the produced material.
The main goal of this paper is tool vibration and its parameters influence
examination and determination of their optimal values. Parameters are
measured and recorded with appropriate sensors, A/D converter and personal
computer. Measurement results are calibrated, saved on computer disk and
statistically analyzed. The paper contain the result of this work.

1. Introduction

Today quality raw wood materials are present in smaller quantities
and requirements for better utilization of those materials are more and
more present. To fulfill those requirements parameters and processes which
influence the degree of utilization have to be identified and optimize.

Primar wood cutting is a process where waste of material is the
greatest in quantity. The main purpose of this process is to obtain semi -
final wood products like planks, bars and shavings. Planks and bars are
more significant products and optimization of primar wood cutting parameters for producing their greater quantities is very important. Otherwise
for final products the quality of semi - final products is very important

too. Quality of semi - final products is also influenced from primar wood cutting parameters.

A lot of parameters influence quantity and quality of primar wood cutting. One of the most important parameter is tool vibrations. Vibrations are result of cutting resistance which is affected from: the speed of log movement, cutting high, tool working time, the speed of tool, etc.

The main goal of this report is parameters influence examination and determination of their optimal values. Our next step is development of an expert system for on line parameter control with micro computer.

2. Log bandsaw vibrations

In the primar wood cutting the most frequently used are log bandsaws which enable different cutting high without blade changing. Tool vibrations produce planks and bars thickness variation and larger quantities of shavings. Increased cutting resistance in this case make tool more dull, produce undesirable thermal effects and tool steel changings.[3] The final results of the extreme tool vibrations besides poor cutting performance lead to catastrophic blade and tool failures.

Luis-Brenta 1400 log bendsaw was chosen for parameter measurement and optimization. Tool vibrations, transporting velocity, tool working time and cutting high were measured. Tool vibrations were measured by an electromagnetic sensor, transporting velocity was measured with current generator and cutting high manually. Measured data are recorded with A/D convertor and IBM compatible micro computer (Fig. 1). Appropriate software for tool working time measurement, data recording and data analysis was developed.

Fagus and pinus logs were chosen for examination. All logs were with similar characteristics according to quality class, log diametar, humidity etc. Data were recorded for 5 fagus and 5 pinus logs for the active time of 32 minutes with data sample rate of 50/sec. Cutting high was devided in 7 classes and entered manually through the keyboard.

Data were statistically analised. All data were devided in to several classes. Analyses were done for 5 classes for transporting velocity, 7 classes for cutting high and 4 classes for tool working time. For each class cutting high and transporting velocity all statistical parameters, like mean value, mean value errors, standard deviation, standard deviation errors, etc. are calculated. Regresion analysis of obtained results was performed and relation between tool vibrations and the other parameters is obtained.

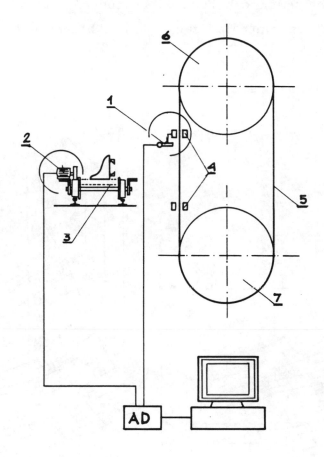

Fig_1. A global schema of harware connection of the system

  1. electromagnetic sensor
  2. smaller generator
  3. transporting car
  4. tool guides
  5. tool
  6. top pulley
  7. lower pulley

On fig. 2 and fig. 3 relation between tool vibration and the other parameters are shown. Results for fagus log are presented on fig. 2. Fig. 3 represents results for pinus log. The graphics are made for first and last four minutes of the tool working time.

Gpaphics show that vibration amplitudes increase with cutting high, transporting velocity and tool working time increasing. The above conclusion lead to an idea to control vibration amplitudes with parameters shanging. The cutting high is a parameter which is constant to each cut. So we can't control tool vibration with cutting high variation for given cut. The tool working time can't be changed too. So transporting velocity is the only parameter which can be used to control vibration amplitudes. Transporting velocity is controled manually at this moment.

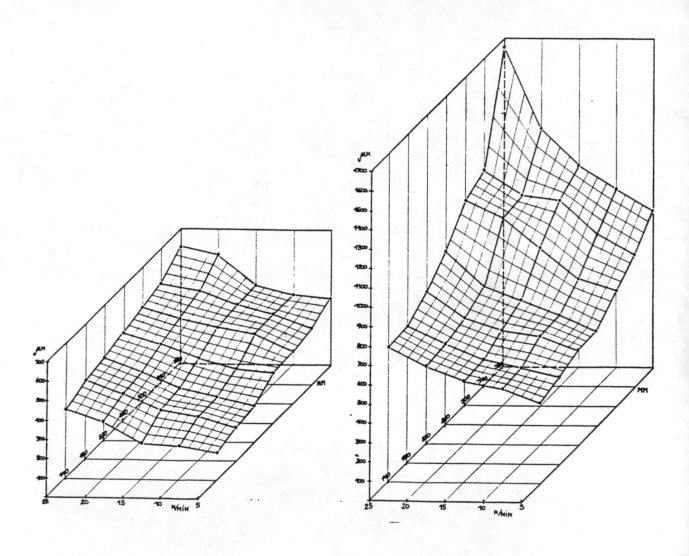

Fig_2. Relation between main values of tool vibration and the other parameters for fabus log

Manuall control of the parameter lead to a lot of errors in the primar cutting process. Errors could be eliminated by comptuter control of
transporting velocity. So an expert system for primar cutting process control is under develoment. The main purpose of the system is to provide
maximum transporting velocity for optimal tool vibration. An expert system
application in primar cutting process offer better utilization of material, tools and machines. The hardware solution of the system and the
software for collecting process data are developed. The software for transporting velocity control is under development.

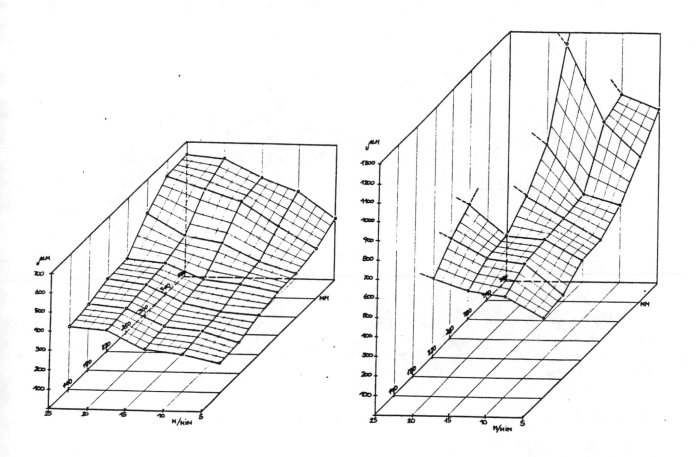

Fig_3. Relation between main values of tool vibration and
the other parameters for pinus log

3. Conclusion

Parameters influence examination and determination of their optimal values are presented in this paper. Connection between process parameters is shown graphically. Graphics show that tool vibration is in direct connection with depending by transporting velocity, tool working time and cutting high. Amplitude control by transporting velocity is discused and conclusions for process control possibilities by personal computer are presented. Our next step is development of an expert system for on line parameter control with micro computer.

Band saw vibration belongs to a class of vibration problems, called axially moving material vibrations. Proposed expert system can be used to control any dynamic system with that class of vibrations with according use of instruments.

REFERENCES

[1] Breznjak, M., Moen, K., 1972, "On the Lateral Movement on the Bandsow blade Under Various Sawing Conditions", Norsk Treteknisk, Inst. Rep., No.46.

[2] Hutton, S.G., 1991, "The Dynamics of Circular Saw blades", Holz als Roh- und Werkstoff 49, 105-110.

[3] Ulsou, A.G., Mote, C.D., Szymani, R., 1978, "Principal Developments in Band Saw Vibration and Stability Research", Holz als Roh- und Werkstoff 36, 273-280.

[4] Klincarov, R., 1978, "Zavisnosta na vibriranjeto na piloviot list i kvalitetot na rezenjeto kaj lentovidna pila-trupcara od visinata na rezenjeto i debelinata na sticite", Sumarski pregled 3-4, 43-58.

[5] Pahlitzsch, G., Puttkammer, K., 1975, "Schnittversuche beim Bandsagen", Holz als Roh- und Werkstoff 33, 181-186.

[6] Hribar, J., 1968, "Sile na alatu za obradu drveta", IV Savetovanje proizvodnog masinstva, Saraevo.

Communication in Low-Cost CIM Systems
A New Approach

N. Girsule and R. Probst

Institute of Handling Devices and Robotics

University of Technology, Vienna, Austria

Abstract: Computer Integrated Manufacturing is based on computer communication. Data exchange is done by connecting workstations, manufacturing cells, machine tools, robots, etc. by networks. Many protocols and standards are defined to ensure safety and reliability of communication. Usually one problem arises in CIM: software costs explode, because CIM solutions almost are specialised programs for one distinct environment. A second problem in automation are the hardware requirements for CIM-software. Software developers assume, that appropriate hardware is present. Many software packages are available only for workstations and mainframes, which are much more expensive than simple PCs.

In most cases small or medium sized enterprises do not have the financial power to buy large computers and sophisticated software packages. They do not really need all the options and processing speed of these solutions. What they need are flexible, small programs, which can be adapted and supplemented easily and fast without external help.

This paper presents an unusual approach to organise communication in small CIM systems: Microsoft Windows 3.x Dynamic Data Exchange (DDE) combined with NetBIOS handle all data exchange and sessions. Both systems use the client/server architecture and work together properly. MS-Windows is in use world-wide. NetBIOS is available for most of the common networks today. Using a small MS-Windows application for DDE-routing full DDE-communication is possible. For external databases or external LANs with different protocols some gateway applications can be designed, which are used as database or communication servers. Users can easily program new applications by hand of the various, really simple programming tools for MS-Windows.

Keywords: CIM, Computer communication, Distributed databases, Networks

INTRODUCTION

CIM as Computer Integrated Manufacturing needs computer communication. Central databases are accessed from various applications. The main groups of CIM-components generate and consume a lot of data which must be interchanged between different applications running on different computers:

⇐ PPS-systems design workplans, manufacturing orders and tool or material reservations.

⊗ CAD-stations create large drawings and piece-lists which are used in other modules.

⊗ CAP uses drawings and technological data to develop work plans, manufacturing and assembly support tools.

⊗ CAM consumes work plans, NC-programs, and production data as manufacturing tools.

⊗ CAQ designs and uses various data from CAD, CAP, and CAM for quality management.

⊗ OCA determines the current state of production and reports it to the controlling system.

⊗ CAO uses drawings and statistical information from CAD, CAM, and OCA to generate reports, public relations material, employee remuneration base data, etc.

All these data must be transported and controlled by communication services.

Today normally Local Area Networks (LAN) are used to connect the different areas of the factories.

Layer	Name	Function
7	Application Layer	Common application service elements like connect, abort, and transfer data and specific application service elements for particular application services
6	Presentation Layer	Transformation of local data representations into those agreed upon for interchange between nodes
5	Session Layer	Connection negotiation, establishment, and release as well as duplex data transfer
4	Transport Layer	Guaranteed delivery, message sequencing, and full error detection
3	Network Layer	Routing of messages from source to destination through intermediate systems
2	Data Link Layer	Media access: protocol to control access to the shared medium (CSMA/CD or Token) Logical link control: connectionless, connection oriented, or immediate acknowledge service
1	Physical Layer	Electrical and mechanical characteristics of the network

Table 1: The ISO/OSI reference model

Various vendors sell various LAN products which use more or less different techniques and standards. Applications use LAN services by specially adapted drivers.

The common problem in integration of autonomous CIM-components are different standards of LAN access and incompatible data formats. This situation results from the history of the CIM-modules. Most of them initially were designed as isolated islands with few external communication. Various vendors offered CIM-modules based on their internal standards. The next step was to connect applications of the same kind. Basical standards for each kind of application were defined. But no respect to other modules within design, planing, and manufacturing was given. When CIM was introduced in a large frame these quasi-standards were used anywhere. Both users and vendors had to invest very much money to override these imcompatibilities between the single islands. At this point it was necessary to create and publish international standards for CIM-integration.

COMMUNICATION STANDARDS

The ISO/OSI Reference Model

The International Standards Organisation (ISO) introduced the Open Systems Interconnection (OSI) in 1983. The ISO/OSI reference model defines communication standards for between open systems as well as inside closed systems. Closed systems use vendor specific communications protocols similar to the ISO/OSI model. Between these vendor specific protocols exist incompatibilities. Open systems base upon the ISO/OSI model. Interconnection between two open systems is possible without much adaption.

The ISO/OSI model divides the complex communication procedure in 7 layers. Each layer has well defined tasks. While communication between to network consumers each layer is used by the layer above in hierarchy. Between the different layers vertical protocols are defined. Inside the layers peer-to-peer protocols handle the communication between the two systems (horizontal protocols).

The 7 layers can be classified in two general groups:

⊗ layer 1 to 4 realise the transport of data,

⊗ layer 5 to 7 coordinate and process application data.

The tasks of the layers are shown in table 1.

The Manufacturing Automation Protocol (MAP)

In 1980 General Motors started a project to standardize communication between machinery coming from a wide variety of vendors. Many other enterprises were involved in this project. The ISO/OSI reference model was used as a base for communication standards. In 1988 MAP 3.0 was introduced.

Table 2 shows the specification of the ISO/OSI layers in MAP 3.0.

Standards for small and medium sized enterprises

MAP as a common standard for communication in industrial environments is a suitable example to show that any CIM application should conform to the ISO/OSI standards. The Manufacturing Automation Protocol comes from large scaled factories. For small and medium sized enterprises some similar approach should be used. CIM systems in these factories should use as much as possible elements of MAP. But often some financial constraints prevent strict use of all MAP definitions.

Otherwise not all the possibilities of MAP and the subsets of definitions are really used in small and medium sized enterprises. Normally cheaper computers are used, very often working under MS-DOS with some LAN connection to a UNIX system. The authors developed a communication system for this special environment which uses many elements of MAP.

COMMUNICATION IN MS-WINDOWS 3.X AND NETBIOS

The new communication system is based on PCs under MS-DOS which are interconnected by a standard LAN. On top of LAN drivers NetBIOS must be installed. NetBIOS is used by the DDE (Dynamic Data Exchange) communication standard from MS-Windows 3.x.

NetBIOS

NetBIOS was introduced by IBM in 1984 with the IBM PC Network adapter card. This first NetBIOS implementation based on the CSMA/CD access method with IEEE 802.3 Ethernet. Since this time many variations of NetBIOS were developed world-wide. This communication interface is available for most of the commercially available networks using protocols like IPX/SPX, TCP/IP or MAP.

As shown in figure 1 NetBIOS resides between layer 5 (session layer) and layer 6 (presentation layer) of the ISO/OSI reference model. Four categories of commands are supported (table 3). Name support and session support make NetBIOS a

Layer	Name	Function
		Application Program Interfaces (API)
7	Application Layer	ISO-MMS DP 9506
		ISO-FTAM DIS 8571
		ISO ASCE DIS 8649/1-2
		ISO ASCE DIS 8650/1-2
		MAP Directory Services
		MAP Network Management
6	Presentation Layer	ISO Presentation DIS 8823
5	Session Layer	ISO-Session Kernel IS 8327 duplex
4	Transport Layer	ISO-Transport Class 4 IS 8073
3	Network Layer	ISO-Connectionless Internet DIS 8473
2	Data Link Layer	ISO Logical Link Control DIS 8802/2
		Type 1 (for Broadband)
		Type 3 (for Carrierband)
		ISO Token Bus DIS 8802/4
1	Physical Layer	ISO Token Passing Bus DIS 8802/4
		Broadband (10 Mbit/sec) or
		Carrierband (5 Mbit/sec)

Table 2: The ISO/OSI layers in MAP 3.0

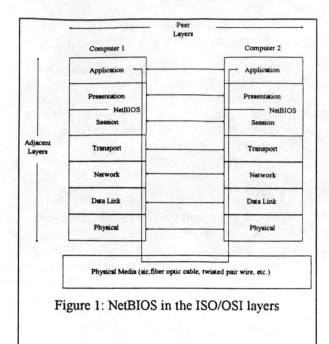

Figure 1: NetBIOS in the ISO/OSI layers

very efficient and flexible tool for creation of distributed systems.

MS-Windows 3.x

Microsoft presented version 3.0 of MS-Windows in 1990. The conception of this version is based on Intels 80386 processor. MS-Windows as an operating system shell offers a graphic user interface of high quality and multitasking. Between the single tasks the communication standard DDE is defined. DDE strictly uses the client/server architecture and includes full protocol and error detection. It is based on the internal message system of MS-Windows. The according messages and their meanings are shown in table 4.

Application, topic, and item are referenced by names.

CONNECTION BETWEEN DDE AND NETBIOS

DDE and NetBIOS both use a client/server architecture and sessions for communication. The

Name Support	Add Name
	Add Group Name
	Delete Name
Datagram Support	Receive Datagram
	Receive Broadcast Datagram
	Send Datagram
	Send Broadcast Datagram
Session Support	Call
	Listen
	Send
	Send No-Ack
	Chain Send
	Chain Send No-Ack
	Receive
	Receive-Any
	Hang Up
	Session Status
General Commands	Reset
	Cancel
	Adapter Status
	Unlink

Table 3: NetBIOS commands

WM_DDE_ACK	General message for acknowledgement. This message uses various parameters to report receipt and processing of other DDE-messages.
WM_DDE_ADVISE	Request to a server application to supply an update for a data item whenever it changes.
WM_DDE_DATA	Send a data item from the client to the server or report availability of data.
WM_DDE_EXECUTE	This message forces a server to execute a command posted as string.
WM_DDE_INITIATE	initiates a DDE conversation for a topic with an application.
WM_DDE_POKE	Request to the server to accept unsolicited data.
WM_DDE_REQUEST	Request to server to provide the value of a data item.
WM_DDE_UNADVISE	Termination of a WM_DDE_ADVISE session.

Table 4: MS-Windows messages related to DDE

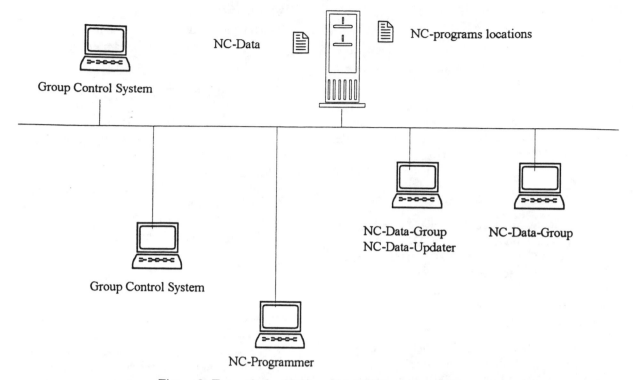

NC-Data

NC-programs locations

Group Control System

NC-Data-Group
NC-Data-Updater

NC-Data-Group

Group Control System

NC-Programmer

Figure 2: Example for DDE and NetBIOS communication

MS-Windows application NetBWin designed to convert DDE requests and sessions into NetBIOS format enables all other MS-Windows applications to fully use NetBIOS communication facilities. This application handles adding and deleting names and group names, establishes NetBIOS sessions and advises the client about changes or results from the remote system.

Based on NetBWin MS-Windows applications are installed as servers on the network. Each server can solve a special task like updating a small MS-EXCEL database or looking up in reference tables. Clients call data via DDE by sending topic and item to NetBWin. Results are presented on the same way. Depending on traffic at one PC two or more servers can be installed.

Using group names large data traffic at one station can be avoided by splitting up requests to two or more stations of one group. Request handling is done by internal communication between the members of one group. Large files are exchanged only by sending filename and location on the common fileserver. There is no need to send the whole file.

Figure 2 shows a small example of communication in this environment:

⊗ A NC machine tool needs a NC program. The Group Control System requests location and filename of the NC program from the group NC-Data-Group. The request is done by sending a piece number and machine type.

Only valid versions of NC programs can be used.

⊗ In this system two servers are installed for this task. They only have rights to read the central database file. Both servers receive the request and one of them looks up for the requested data in the central database (e.g. MS-EXCEL sheet).

⊗ Location and filename are sent back to the client (GCS). The client now opens the specified NC data file and copies the program into the NC machine tool.

⊗ At the same time the NC programmer changes another NC program at his station. After completition he stores the new file on the fileserver and requests to update location and filename of the new version of this NC program in the same central database as above. This request is sent to the server with the unique name NC-Data-Updater.

⊗ The NC-Data-Updater has rights to write to the database file. After performing the update an acknowledgement is sent back to the client.

⊗ The client receives the acknowledge and terminates the session.

This simple example shows the mechanism behind this communication method.

⊗ Each single task is implemented in a special MS-Windows application. They are referenced

85

by unique or group names via the DDE to NetBIOS interface NetBWin.

⊗ Infrastructure of the Local Area Network is fully used for data storage, data exchange, and transmission.

⊗ Large databases can be splitted up into small and easy to be updated files or spreadsheets. Access can be done by world-wide available programs like MS-EXCEL.

CONCLUSION

Various standards were defined for communication in factory automation. For LANs the ISO/OSI reference model and the MAP 3.0 definitions are the best way to guarantee compatibility and stability. Small and medium sized enterprises sometimes do not have the possibility to realise a complete MAP system.

This paper presents a suitable, cheap and easy to be realised way to fully integrate all single tasks in a manufacturing environment. The approach is to use MS-Windows' DDE communication standard combined with the common NetBIOS definition.

Via NetBIOS a remote-DDE communication is implemented and any MS-Windows application can exchange data with any other application.

REFERENCES

Schwaderer, W.D. (1988). *C Programmer's Guide to NetBIOS*. Howard W. Sams & Company, Indianapolis.

Tangney, B., and D. O'Mahony (1988). *Local Area Networks and Their Applications*. Prentice Hall, Herdfordshire.

Petzold, Ch. (1990). *Programming Windows*. Microsoft Press, Redmond, Washington.

Scholz-Reiter, B. (1991). *CIM-Schnittstellen*. Oldenbourg, München.

Norton, P. and P. Yao (1992). *Windows 3*. Markt & Technik, Haar bei München.

Kopacek, P., N. Girsule, J. Hölzl (1992). A low cost modular CIM concept for small companies. *Proceedings of the IFAC-Symposium on Information Control Problems in Manufacturing Technoloy - INCOM '92*, Toronto.

INCOSOFT

Integrated-Company-Software for small and medium sized Companies

Wolfram Wöß, Christian Gierlinger

Research Institute for Applied Knowledge Processing (FAW), Johannes Kepler University Linz, Castle Hagenberg, A-4232 Hagenberg,
Austria, Net:chg@faw.uni-linz.ac.at, Tel: (+43-7236) 3231/58, Fax: (+43-7236) 3338/30

Abstract. The goal of this paper is to describe the components of our system INCOSOFT, the hard and software requirements, some implementation aspects, the differences between the software-systems of small and medium sized companies and large sized companies and the data model. Programming was done in the 4th generation language SQLWindows, data administration was handled by SQLBase. The CIM-software runs on a PC-LAN with a central fileserver and a database-server.

After the indroduction which focus the differences between software-systems for small and medium sized companies and large sized companies, the first part of the paper is dedicated to the specification of our system INCOSOFT. A description of the components of the system is given in chapter 3. The focus of the following chapter is on the implementation aspect. An overview of the result and further activities is given in the last chapter.

Key Words. CIM, Leistand, SQLBase, SQLWindows, INCOSOFT, Database Application.

1. INTRODUCTION

INCOSOFT (Integrated-Company-Software) is the name of a product, which is the result of a co-operation between a saddlery and the Research Institute for Applied Knowledge Processing at the Johannes Kepler University in Linz. The saddlery consists of about 30 employees. It produces saddles including all the fittings.

Since 1987 the saddlery has tried to computerized the company. But there was actually no success. The development of an Austrian software-company caused more troubles than help for the company. The fees raised monthly for the development and adaptation of the system which should specially be produced for the saddlery. Due to the exploding costs in contrast to the never increasing engineering progress of the system, the experience of the saddlery and of many small or medium sized companies is, that the establishment of a computer in such an unprofessional way is not profitable. Therefore the saddlery contacted the Research Institute FAW at the University of Linz for help. Together we tried to search for a solution, appropriate for small and medium sized companies.

First we tried to find a standard software-product, with the following requirements:

- a relational database as the basis of the whole system
- a client-server architecture for data transfer
- a server-PC-network (LAN) is supported
- a graphical user interface with mouse-support

During two years we have not found any software product which fulfills these requirements. There exists some systems for large companies but the installation and adaptaion of these systems in small and medium companies is impossible. The reasons are that in small and medium sized companies:

- normally no EDP department exists and the budget for hardware and software is rather small
- the organisation in the company is confused and often managed with practical knowledge
- data for control is not available due to the confused organisation. An exact calculation of the products is impossible.
- the orders of the clients must be treated very flexible because small and medium sized companies normaly don´t produce in series

- the employees are not skilled in handling a computer environment and therefore the requirement to the system is an user-friendly interface.

Due to the lack of an adaptable existing system we have started to develop our own software-system INCOSOFT.

2. SPECIFICATION OF INCOSOFT

2.1. Hardware

The basis of the hardware-structure of INCOSOFT is a local area network, including a fileserver, a database server and four workstations. The additional hardware consists of a laser printer, three pin printers, a streamer and an electronic control clock. The energy comes through an uninterruptible power supply.

The network and the database services are running on two different servers. This fact has two important advantages. The runtime of applications dramatically decreased and the security of data increases, when it is possible to use two servers. In this case the fileserver services all "normal" applications and the database server does all the operations on databases.

A short example should explain, the increasing security of data: Every complex application includes fatal errors and bugs. The outcome of such an error may be, that the workstation or the fileserver or both goes down. If database-files are still open and the database-services are running on the fileserver, it could be possible, that data gets lost. When you separate the database server from the fileserver the database will not go down if an application error occurs and the data is safe. The damage of data has its reason in the use of cache-managers which store a great part of the data in the main storage of the computer. This data gets lost, when the computers goes down.

The saddlery has a network licence to increase the number of workstations up to ten. It should not be a problem to support every important working place of the company with its own workstation.

The electronic control clock is connected with one workstation. This makes it possible to transmit data from the control clock to the workstation and from the workstation to the database. The clock consists of:

- a display with the actual time and text which includes the operation of the employee and other special information on it

- a keyboard with function-keys and

- a barcode-reader which can read the information of every production-step from the production-order-sheets.

2.2. Software

The operating system of INCOSOFT is based on MS-DOS (Microsoft Corporation, 1991). The reason, why we have decided to use the operation system DOS, has not been absolutly the quality of this system, but the great numbers of software products which can be installed under DOS argue for it. We use Novell (Novell, Incorporated, 1991a; Novell, Incorporated, 1991b) as network software and SQLBase from Gupta Technologies (Ring, Bruce, 1990) as relational database management system. The communication in the network uses the NetBIOS-standard (Novell, Incorporated, 1991b), which makes the system compatible with most software-systems. The compatibility is one of our main principles. Therefore we decided to use frequented software-systems, such as DOS or Novell NetWare.

In the last two years we made the experience, that only graphical user interfaces achive the special demands of all the end-users. In many cases such users never work with a computer before or they have only little experiences with the computer or they are unskilled. Our solution of that problem was the use of the graphical user interface Windows 3.1 (Microsoft Corporation, 1992), manufactured by the Microsoft Corporation. Today in the saddlery only the software for the streamer runs with a conventional user interface. All other software-components are based on Microsoft Windows.

The development-system of the CIM-software consists of

- the Microsoft C-compiler 6.0 (Microsoft Corporation, 1990b)

- the Microsoft Windows Software Development Kit (SDK) (Microsoft Corporation, 1990a)

- SQLWindows 2.0 (Klein, Karen, 1990), a 4th generation programming language.

One important characteristic of SQLWindows is, that the programs developed with it can run in a multi-user- or single-user-system, without changing their sourcecode. In the single-user-system (see fig. 1) Microsoft Windows is the basis of the whole system. SQLWindows is the executable that runs the SQLWindows application and DBWindow is a DOS single-user database server program for Microsoft Windows applications. To connect the application with the database, there are two dynamic link libraries, SQLAPIW and SQLDBW.

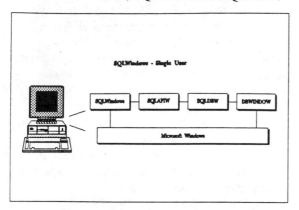

Fig. 1. SQLWindows used in a single-user-system.

In the multi-user-system (see fig. 2) the software-structure in a workstation is very similar to the single-user-system. But in this environment, DBWindow does not fulfill the services of a single database, it connects to the database on the database server, which is a separate node on the network. The additional dynamic link library SQLNBIOW specifies to use NetBIOS to pass messages between a remote client application and the server.

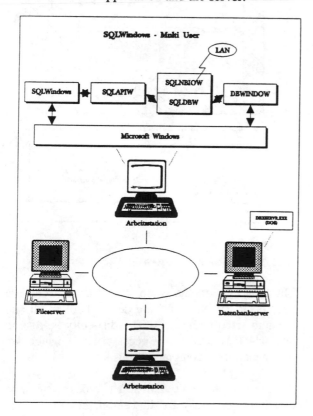

Fig. 2. SQLWindows used in a multi-user-system.

If a workstation should provide a connection to the database for a database-application, the software-structure has to be implemented as described before. On the database server runs a special server application program under DOS called DBXSERVR.EXE. All non-database-transactions are still serviced by the traditional fileserver.

There is no additional programming effort necessary to distinguish between the single-user- and the multi-user-system. This is the function of the three dynamic link libraries (SQLAPIW, SQLDBW, SQLNBIOW) described before.

2.3. Integration aspects

One important aim of CIM is, the integration of all necessary components to a global system which is required to achive computer integrated manifactuing in a company. The user interface Microsoft Windows of the CIM-software, as the basis of the whole system, makes it possible to integrate standard-software with non-standard-software. The outcome is a high-level standard and a great compatibility with many hardware-systems.

The use of standard-software is very important for a small and medium sized company, because these products are components of the global system, which are well-tested, with a high functionality and in many cases with a low price in relation to the functionality. Under the operation-system-extension Windows our saddlery uses four standard-software products:
- Microsoft *Excel* (spreadsheet),
- Microsoft *WinWord* (word processing),
- Corel Systems *CorelDraw* (drawing program)
- *CAD-system*.

This standard-software is full compatible in the outfit and handling with INCOSOFT.

2.4. User Interface

The user interface of INCOSOFT is user friendly. It fulfils the IBM-CUA/SAA standard and has full mouse-support. It was developed not only for skilled users but also for unskilled users. .

Our demand was, to design a user interface, so that the handling with the CIM-software makes less expenditure for the employees of the firm. This seems to be important, because a small or medium sized company is not as big that the employment of a specialist for the handling of the CIM-software would be justified from the financial point of view.

Nevertheless the CIM-software has to support the necessary functions of a firm. Therefore it is not possible to leave such functions out. Our solution of that problem was the special structure of the user interface, where the most important functions are placed in the main-window of each component and the less important functions can be reached with pull-down menus, push-button and so on.

3. COMPONENTS OF THE SYSTEM

INCOSOFT consist of all well-known parts of a CIM-architecture (see fig. 3). These are the
- resource management
- financial accounting
- word-processing
- spreadsheet
- computer aided design
- staff management statistics
- order-administration
- delivery
- price-list management
- invoice
- administration of customs
- store administration
- controlling and calculation
- shop-floor-monitoring
- product-planning-system including a graphical leitstand

The basis of the whole system is a relational database, so that every workstation can reach the latest versions of data and can use all integrated components of the CIM-architecture.

The administration of business-partners, commodities and employees is yet implemented. For the components financial accounting, word-processing, spreadsheet and computer aided design we have found appropriate standard software. Now we are developing a shop-floor-monitoring especially for small and medium sized companies. In a further project we are going to develop a product-planning-system and the graphical leitstand.

4. IMPLEMENTATION ASPECTS

After the analysis of the requirements was complete we use the Entity Relationship Diagramm (Chen P. P. 1976) to develop the data model for the entire company. The result was 74 entities and 59 relationships. For the dynamic representation of the universe of discourse we use the Structured Analysis and Design Technique. The entire data model was transfered into a relational database and the functions of the SA-Methode are realized with a 4th generation language.

The relational database SQLBase from Gupta Technologies is the basis of the whole system and a client-server architecture is the underlying concept of the database. As already mentioned before, also a server-PC-network (Local Area Network) is supported.

The most parts of the CIM-software is implemented in the 4th generation language SQLWindows. The object-oriented design of the software-system makes it easy to adapt the sourcecode to special demands of other companies.

A few smaller parts of the system are implemented in Microsoft C. For instance the functions for the data-transmission to and from the control clock. These and other functions are collected in special dynamic link libraries, so that they can be used by many other client-applications.

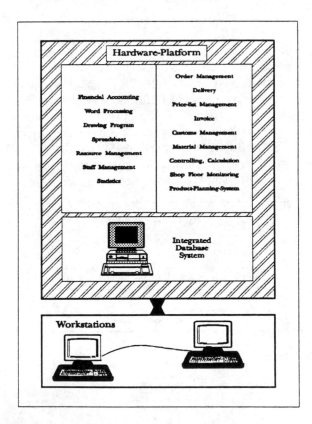

Fig. 3. The components of INCOSOFT.

The concept of the shop floor monitoring are developed especially for the demands of small and medium sized companies. The difficulty is, to find out, which functions are necessary and which are not needed for a company of this size.

We are planning to establish a barcode system, so that it is very easy for the employees to use the control clock. The customer-orders are collected to production-orders. For every production-order the steps of the production are printed on a special sheet. At the beginning and at the end of every production-step the employee has to transmit special data (begin and end time of the step, flow time etc.) to the datahase. For a correct transmission of this data to the database we use the control clock in combination with the barcode system. Data entering in this way was very easy for the employees by using the barcode.

5. CONCLUSION

We have been trying to introduce our System INCOSOFT. We have discussed the specification of the system INCOSOFT especially the hardware and software specification. We have shown the user interface and the some implementation aspects. We have sketched the differences between the installation of software-systems in large sized companies and small or medium sized companies.

The main advantage of our system is that the user is able to use INCOSOFT togehter with fully integrated standard software products. Furthermore the system consists of an uniform standardized user-friendly interface with full mouse support.

We continue to implement INCOSOFT as described in the paper. Further developing work has to be done in this area especially in the field leitstand in combination with relation databases.

6. REFERENCES

Chen P. P. (1976). The Entity Relationship Model: Towards a unified view of data. ACM TODS, 1st edition, Volume 1, 9-36.

Gardarin G.: Extending a Relational DBMS towards KBMS: A First Approach, Workshop on Knowledge Base Management Systems, Creta, June 1985

Gierlinger Ch., A M. Toja, R. R. Wagner (1988). Die Entwicklung im Bereich der Datenbanken als Prozeß der Wissensverlagerung. In Statistik, Informatik und Ökonomie, Springer Verlag, Berlin Heidelberg, 1988.

Gierlinger Ch., P. Andlinger; (1991a). Objektorientierte Datenbanken als Grundlage für Offene Systeme. Conf. Proc. Arbeitsgemeinschaft für Datenverarbeitung, Wien, November 1991.

Gierlinger Ch., W. Retschitzegger, R. R. Wagner (1991b). Datenbanken für CIM. In Tagungsband

des 7. Österreichischen Automatisierungstages, ed. P. Kopacek, Wien, Oktober 1991

Klein, Karen (1990). SQLWindows, The Graphical Database Application Development System, Technical Reference Manual. Gupta Technologies, Incorporated, Kanada.

Meynert, J. (1990): "PPS*STAR/L1 - ein PPS-System für mittelständische Einzel- und Auftragsfertiger", CIM Management 1/90.

Microsoft Corporation (1990a). Microsoft Windows Software Development Kit, Guide to Programming. Microsoft Corporation, USA.

Microsoft Corporation (1990b). Microsoft C - Advanced Programming Techniques for MS OS/2 and MS-DOS Operating Systems. Microsoft Corporation, USA.

Microsoft Corporation (1991). Microsoft MS-DOS Benutzerhandbuch und Referenz für das Betriebssystem MS-DOS Version 5.0. Microsoft Corporation, Irland.

Microsoft Corporation (1992). Microsoft Windows Benutzerhandbuch Version 3.1. Microsoft Corporation, Irland.

Novell, Incorporated (1991a). Novell NetWare Version 3.11, System Administration. Novell, Incorporated, USA.

Novell, Incorporated (1991b). Novell NetWare Version 3.11, Installation. Novell, Incorporated, USA.

Ring, Bruce (1990). SQLBase/SQLTalk, Language Reference Manual. Gupta Technologies, Incorporated, Kanada.

Scheer A.-W.. (1987) "CIM, Der Computergesteuerte Industriebetrieb", Springer Verlag Heidelberg-Berlin-New York-Paris-Tokyo, 2. Auflage, BRD 1987.

Wöß, Wolfram (1992). CIM-KM Computer Integrated Manufacturing für Klein- und Mittelbetriebe, Stammdatenverwaltung für Geschäftspartner, Systemverwaltung, Warengruppen, Artikel und Mitarbeiter, Wissenschaftlicher Bericht im Rahmen eines Programmierprojektes, Forschungsinstitut für Anwendungsorientierte Wissensverarbeitung, Abteilung für Datenbank- und Expertensysteme, Johannes Kepler Universität Linz, 1992.

INTRODUCTION OF FLEXIBLE AUTOMATED ASSEMBLY INTO SMALL ENTERPRISES IN SLOVENIA

D. NOE

Faculty of Mechanical Engineering, University of Ljubljana, Slovenia

Abstract. The paper presents the state of the art of the assembly process in Slovenian enterprises. Two different approaches to introduction of flexible assembly systems are presented. In the first example a new assembly system was developed for a new product using new assembly technology. In the second one the integration of logistics and assembly systems is discussed to ensure the necessary flexibility.

Key Words. Assembly; flexible assembly; logistic; computer controlled assembly;

1. INTRODUCTION

By the rule, assembly process accounts between 40 and 60 % of total production time and is in most cases carried out manually. Increasing market demands for better flexibility and quality, is forcing the development and introduction of new assembly systems. As the 50% of Slovenian enterprises are so called "small enterprises" and the introduction of the totally automated assembly systems is very capital intensive, progress in the future will be performed in steps. One way of the assembly rationalization is the introduction of new flexible assembly systems for new products. A successful connection of manual and automated assembly is also appropriate for small enterprises. Some examples of solutions already implemented in the production process, or planned in the future, are presented in this paper.

2. SURVEY

In 1991 a study of assembly systems in Slovenian enterprises was made, to get a survey of the state of the art of assembly in Slovenia (Noe 1991). The aim of the study was to obtain the fundamental data for establishing the strategy for research work, which should improve the productivity of enterprises and gradually introduce computer aided assembly (CAA), as a part of computer integrated manufacturing (CIM).

Taking into account that the Slovenian machine building, electromechanical and automotive industries have 75.000 employees and about 20%, are working in assembly and that the average number of employees in enterprises is about 200, it can be

concluded that most of Slovenian enterprises are so called "small enterprises".

The obtained data in the study show that the production is in 50% small series, 30% large series, 10% individual and 10% mass production. A conclusion drawn from this fact is that only a small part of products are suitable for the conventional automated assembly.

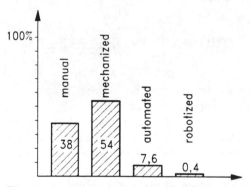

Fig. 1. Automation grade of assembly systems

The same conclusion is obtained also from the analysis of assembly operation stations. Only 7.6% assembly operation stations are automated, 54% of them are mechanized, 38% of them are manual, using hand tools only and merely 0.4% of assembly operation stations are robotised (see Fig.1). Robots in the assembly process are a real rarity and are used mostly in the health hazardous assembly operations. A self-contained assembly cell has not yet been introduced into production. Robots are placed at the assembly lines and are accomplishing predefined simple tasks.

Similarly undeveloped is the automation of transport systems between the assembly operation stations. A flexible automated transport system is connecting only 5.5% of assembly operating stations, 41% of stations are connected by assembly belts and the rest of assembly operation stations are grouped or not properly connected.

The surveyed sample of enterprises was extremely heterogeneous. It included from home appliances industry, automotive industry, down to production of small metallic products. Therefore the observed products are extremely different.

In the analysis of possibilities for the introduction of automated assembly, the difficulties were stated and classified. The most important obstacle to introducing automated assembly was considered to be small series production and unsuitable design for automated assembly. In the second place was the multitude of different variants of products and on the third place the unsuitable quality of assembly parts.

An important issue of the study was the answer to the question where the equipment for assembly automation is obtained. In 57% the answer was that the equipment was produced by the company itself, 27% of the equipment was imported and the rest was bought on the home market. The fact that most of the equipment is designed and produced locally is confirmed by data of personal designated for this purpose. In small enterprises such a group consists of three and in bigger ones, up to ten experts with high degree of education. According to this, a fast growth of automation and introduction of flexible assembly systems is to be expected in a better economical situation in the future.

3. INTRODUCTION OF FLEXIBLE AUTOMATION

It can be assumed that the general assumptions about the rationalization of the assembly process are also valid for Slovenia (Lotter 1992). Slovenian experts point out the following goals to be achieved by introducing flexibility into the assembly process:

- assurance of integral product quality,
- decreasing the assembly and/or production costs
- shortening of production time
- increasing the degree of automatization
- ability for fast response to market changes
- fast introduction of new products

To achieve these goals computerized data control has to be introduced, suitable production equipment has to be designed, organization of enterprises has to be improved, and the personal in all production levels has to be better educated and trained.

A specific problem of Slovenia, as well as of all post communist economies, is the lack of production costs acquisition and evaluation. There is very little or no experience in realistic production costs determination, and therefore the most important data for the decision of introduction of flexible assembly systems, economics is missing.

Next we would like to present two cases of introducing flexibility into the assembly process. In the first case a flexible assembly system was introduced for a new product, and the second case is a step by step introduction of automatic assembly in connection with automation of logistics.

3.1. New products - new assembly technology

A company producing headlights for different European car producers is introducing automation into the assembly process. Considering their long experience and the analysis of the assembly process, they found it economically to introduce a flexible assembly system for the new products. The life time for car products is short. The development time for new car types, and also the subassembled parts, is about two to three years. The development time for assembly systems and their introduction into production is restricted to maximum 6 months. A uniform high quality and reliable delivery is absolutely necessary. All these requirements force the car components producers to introduce automation into their production.

In the above mentioned company, automation of assembly is made in three steps. The first step was made in 1988 when an automated line for glue application in the headlamp assembly was introduced. The second step is the introduction of a flexible assembly system in 1992 and the third step is planned for 1995 when the computer controlled system for production and assembly of headlights in terms of computer integrated manufacturing will have to be developed.

3.1.1. Assembly line for glue application

The assembly process for headlamp P-204 consists of 10 assembly steps (see Fig. 2). The assembly process is divided into three phases: glue application, drying and final assembly. Automation of glue application was necessary because of toxic glue and high health hazard for the involved workers, desired high quality of adhesive junction and the necessity of heating before and after glue application. 8 of 10 assembly operations are accomplished on the assembly line. The assembly line consists of 2 manual and 6 automatic assembly stations which are connected by a pallet transporting system. Two further operations are made separately.

Pallets are transported from one assembly station to another by a chain transporter. Glue application is made by a programmable manipulator and the gluing shape is defined by a mould. The system is controlled by a programmable controller.

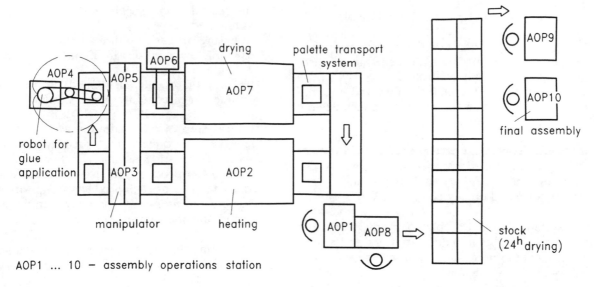

Fig. 2. The headlamp P-204 assembly system

The original pace time of the system was 2 minutes. By optimization of the movements it was possible to shorten the pace time to 1.5 min. Modular design enables the system to be used for assembling three different headlight types.

3.1.2. Flexible assembly for the headlight VW92

The company's own experience and the customers' demands initiated the development of the flexible assembly system with maximal possible automation level of assembly and control operations. An assembly line was devolved, with manual assembly stations included where automatization was not possible or not economical (see Fig. 3). The system enables the assembly of the left or right side headlight as well as both variants with and without adapter in random sequence. The pace of the line is 20 seconds. The pace is determined by the robotized station for glue application. The assembly line includes two SCARA robots, four automated assembly stations, pneumatic manipulator for lenses placing, 6 manual assembly stations and two automatic control stations. All the automatic and manual stations are connected by a palette transport system. Palettes are coded. All the stations are individually controlled. The palette transport system is controlled by a programmable controller. The two automatic control stations are controlling the relative position of paraboloide to the frame and tightness.

The research work concerning the introduction of flexible assembly system was supported by the Ministry for Science and Technology of Slovenia.

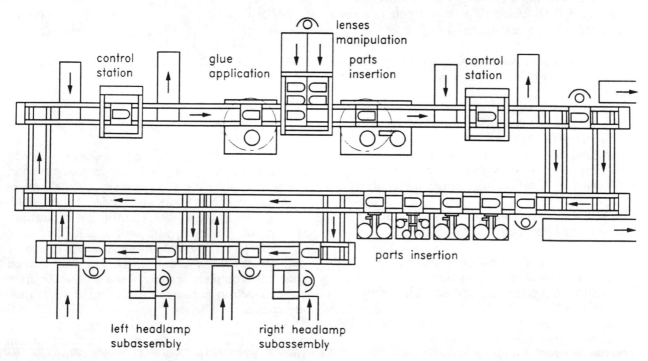

Fig. 3. Flexible assembly system for headlamp VW92

3.1.3. Computer controlled assembly

The gathered experience and future development considerations pointed out the necessity of introducing of a computer controlled assembly. In the proposed concept, the material flow and information flow are considered. The system includes the following components:

- computer controlled parts manufacturing (CCM)
- computer controlled material flow - logistics(CCL)
- computer controlled assembly (CCA) and
- computer aided quality control (CAQ)

The system should be able to produce different types of headlights in many variations and ensure the assembly in accordance with customers timing and quality requirements. The concept ensures optimal material flow throughout the production. The intermediate stock for parts is no more necessary, production of bulky parts (frame and paraboloides) is just in time for assembly without intermediate stock, production of some small parts is placed besides the assembly line, the storage time for the purchased parts is minimized. Transport for parts and finished products is automatic by AGV. The system includes a minimal finished products stock.

3.2. Integration of logistic and assembly systems

A company specialized for production of ball valves, safety valves and components for central heating is producing 49 different products in three to eight variants. Assembly system is manual, using a few mechanized assembly units. Assembly is organized in 25 individual assembly stations, three group assembly stations, one automatic assembly unit and two quality control stations. The parts are transported from stock to assembly and finished products back to stock again.

The jobs are issued sporadically, according to customers orders which are random in qualities and time.

The problems of the existing assembly system are:
- poor flexibility
- long time from order reception to delivery
- high capital costs for subassembled products
- difficult identification of production failures

It was decided to rationalize the production to achieve following goals:

- to ensure the desired quality and tracking the order throughout the production.
- to ensure controlled material flow throughout the factory
- to integrate the assembly process with the planning department and other activities in the factory by introducing an integral computerized information system
- assembly is going to remain manual in the near

future, only some specific stations will be automated
- to ensure an easier future automation by assembly oriented design of parts and products.

The solution of the mentioned problems is the introduction of computer aided production.

The first step of rationalization was an analysis of the existing products and assembly system. The concept of existing assembly system is based on grouping the products according to their assembly requirements. For each products group with similar assembly requirements, assembly stations are organized and the workers trained.

The crucial part of the new production concept is integration of assembly and logistic systems (Schmid 1990).

A new storage and transport system has to be developed an realized (see Fig. 4). The integral transport system has to accomplish transport of parts from workshop to stock, transport of locally produced and purchased parts to assembly stations, transport between individual assembly stations, transport from assembly to quality control stations and back to stock, as well as transport from stock to packaging and delivery.

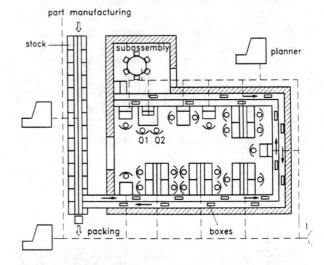

Fig. 4. Integration of assembly and logistic systems

As the maximal weight of products does not exceed 3 kg, a transport system with standard palettes and boxes with a combination of belt and reel conveyers was chosen. Boxes are coded by magnetic coding system. Assembly, control and production stations are equipped with code reading and code writing devices. The code includes all the necessary data about the contents. This kind of identification enables a complete control of the production from the production of parts through assembly and quality control to delivery.

Assembly stations are organized and equipped in such a way that assembly of different products of the same group is possible. The workers are trained for assembling all the products of the group. If necessary the same product can be assembled on more assembly stations. This enables a faster response to larger orders. The task of the planning department is to coordinate the issue of jobs with orders and individual assembly stations. The maximal number of assembly stations for the individual group of products is determined by the analysis of last year orders and prognosis for the next year. Assembly stations are equipped with modular assembly equipment. Special assembly tools specific for individual products are stored in stock and delivered to assembly station together with parts. This system requires a maximal number of assembly tools to be on stock. However, because of modular concept the costs are considered not to be too high. The assembly personal has to be able for random and fast changes of assembly procedure. The system is considered to be capable for several changes of assembly procedures in one day.

Most of the products involve a subassembly consisting of an axis with washers. This subassembly, however in different dimensions, occurs in large quantities. Therefore the operation of placing the washers on the axis, is automated.

The system will also include automatic assembly of valves produced in large quantities, which makes the development of an automatic device economical.

The computerized data system integrates stock departments, assembly, quality control, production, and planning department. All the necessary data about products or semiproducts are coded on the magnetic code on the transport box. This data is automatically fed into the computerized data system when the box enters the stock. At that time a worker verifies if the data in the computer is correct.

Data base necessary for functioning the system consists of following data:

- assembly technology data
- equipment data
- availability of assembly and control stations
- stock data
- parts and subassembly data
- orders data

The system is controlled by planning department, which issues the jobs and also provides the data about changes in the assembly system, such as new products, new assembly procedures etc. Information is first transferred to the stock. A stock worker dispatches the parts and tools to the assembly stations. All dispatching data is available from computer.

Relevant data from the assembly control system is currently fed to the company's central information system.

The shown concept is going to be realized next year as a pilot project. The research work is supported by the Ministry for Science and Technology of Slovenia.

4. CONCLUSIONS

The analysis shows that a planned and general introduction of assembly automation is not yet taking place. But there is a great interest for new solutions in enterprises and some of them, specially those that are involved in the international bossiness, gradually introduce new production methods.

The role of the Laboratory for Handling and Assembly on the Faculty of Mechanical Engineering in this process is double. In mutual projects supported by the Ministry for Science and Technology, the laboratory is cooperating with enterprises solving the actual problems and developing and introducing new assembly systems and computer controlled assembly.

The second important task of the Laboratory is education of the experts coming from enterprises in the fields of computer controlled assembly systems, quality assurance in the assembly process, economical justification of new solutions and assembly - oriented design of products and parts. As the research showed that the most equipment is made locally, the Laboratory is developing project methods for planning the flexible assembly systems.

Study of the development of flexible assembly systems in Slovenia pointed out the following facts:

- the enterprises are forced to introduce computerized control over the assembly process and entire production to ensure the desired quality and fast response to customers orders,
- rationalization and introduction of flexible assembly is rational for new products which are produced in large enough series,
- modernization of assembly for fading out products is not rational,
- manual assembly for producing small series products with a long production life can be rationalized by the introduction of transport automation and computerized control or integration of assembly and logistics.

5. REFERENCES

Lotter B. (1992).*Flexible Montage*, Rechnerintegrirte Konstruktion und Produktion 8, VDI Verlag GmbH, Düsseldorf 1992.

Noe D. (1991). *Production rationalization by introducing of Flexible Handling and Assembly Systems,* (Report in Slovene) 1991.

Schmid J. (1990).*Integration of Assembly and Logistic* System (IMOLOS-EU 412), 2nd FAMOS Advanced Course on Flexible Automated Assembly, Venice 1990.

INTEGRATION OF CAD/CAM AND PRODUCTION CONTROL FOR SHEET METAL COMPONENT MANUFACTURING ON PC

A. Witters

Wetenschappelijk en Technisch Centrum voor de Metaalverwerkende nijverheid (W.T.C.M),
Celestijnenlaan 300 C, B-3001 Leuven, Belgium

Abstract: The paper deals with a computer integrated manufacturing system for Sheet Metal components, developed in the frame of the Brite-P-2406 project. The implemented prototypes handles the entire information flow from design to final product and demonstrates the reduction of delivery times and costs using such an integrated system. Considering the large investments regarding hardware and software demands and the high degree of automation, the use of the system is not really suitable for SME's. Therefore, the key modules of the system, which were originally developed on work station, are currently integrated with existing modules to obtain a powerful low cost solution running on PC.

Key words: CAD/CAM, Production planning and control, Features, Data bases, Sheet Metal

1. INTRODUCTION

Sheet Metal comprises several phases, including the design of the workpiece, process planning and planning by order selection. Tools for supporting those phases are nowadays offered as stand-alone packages. The few exceptions, offering an integrated system, are linked to expensive hardware and software, and require a highly skilled informatician to manage and maintain the system.

The reason is that a complete and optimal automation of each step in Sheet Metal Manufacturing requires high computation times. Consequently using such a highly integrated system can only be justified for huge production volumes, and they are out of the question for small and medium sized companies, typically involved in subcontracting and with no informatics personnel.

Considering the large amount of existing Sheet Metal packages, the development of a completely new system would make no sense. Therefore, the project rather focuses on reusing existing products e.g. CAD, CAM, scheduler etc. These applications are downgraded to the specific needs of smaller companies.

This paper introduces a fully integrated solution for a manufacturing system, in this case applied to Sheet Metal manufacturing, that has been integrated with existing (commercial) modules to obtain a powerful low cost solution running on PC.

2. THE NEED OF AN INTEGRATED MANUFACTURING SYSTEM

Over the years, most approaches to automation of production planning have been characterized by the absence of integration with other products as CAD and MRP systems, sometimes present in the company.

On the other hand, process plans and schedules are generated, long before the actual production. Often the assumptions made during the process planning hardly resemble the actual state of the company at production time, leaving a lot of detailed process planning and scheduling to the machine operator or the shop floor supervisor. This working method has obviously some disadvantages:
* Difficult feedback from production to design

* Long lead times between product design and product delivery
* Non-flexible production system due to sequences and deadlines for orders being fixed a long time in advance

Within the Brite-P-2406 [1] Project, WTCM developed and implemented a Sheet Metal component manufacturing system that integrates functions like design of the workpiece, process planning, planning by order selection, nesting, automatic NC-code generation, shop floor monitoring and control. Taking the real situation of the shop floor into account, one can provide maximum flexibility for optimisation of performance in the actual production. The figure below illustrates the structure of the system.

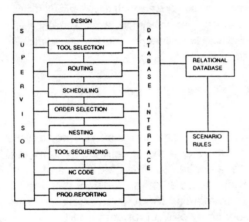

Figure 1: structure of the integrated system

The supervisor module has a double purpose:

* calling standard modules in a certain order to generate NC-code and other information ready for production
* calling modules to react on events occurring on the shopfloor or during NC-code generation

The set-up of the system is defined in "scenario" tables in the database. These scenario's are used to define the functionality of the supervisor.

The modular structure makes it not only easy to add new specialized modules but makes it also possible to adapt the system to the needs of a specific company.

All system modules work on a central relational database and can be called in any desired sequence.

3. THE COMMON TECHNICAL DATABASE

The main goal in automation is not optimizing one production step but passing the existing information from one process to another.

Most existing computer applications are stand-alone packages and are automating one step in the manufacturing process. Each application has his own data model with specific data representation. As no coordination exists among the distinct applications, data are represented in several ways and at several places. Thus the information easily becomes inconsistent. Integration of these applications not only involves "communication" but also "translation" and "interpretation" of different datastructures.

To avoid these gaps in the information flow between the applications, a common technical database was chosen for the distribution of the data among the modules in the system. In the project four essential categories of data have been identified:

Product data: contains the geometrical and technical description of the product.
Raw Material data: contains information about material.
Resource data: contains information about the characteristics and availibility of tools, machines etc.
Process Planning data: contains information regarding routings, order description, scheduling etc.

Each of these categories consists of a fixed and variable part. "Fixed" data are basic elements and don't change eg. the definition of a form feature. "Variable" data consist of combinations of these basic elements and are the result of certain modules eg. the description of a product, an order etc.

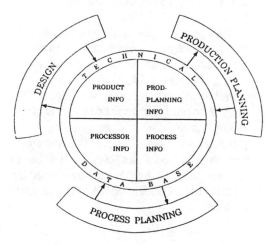

Figure 2: General approach to the distribution of data among several applications

4. FEATURE BASED DESIGN SYSTEM

Integration between design and the other technical functions means that the feature based workpiece description, generated during the design phase, is used as input for the other technical functions. Features are used as communication medium between design and manufacturing systems. The use of features not only facilitates the task of the designer, but also standardises the representation of all data concerning the workpiece.

Several types of classifications can be made for the form feature:

On one hand, a distinction is made between several levels of information. E.g.:
* Difference between the element(s) describing the contour of the workpiece (=perimeter) and the internal holes (=element)
* Difference between the complex element and the basic elements describing each line, arc, corner or notch of the complex element (=component)

On the other hand, a certain hierarchy is included to allow the building up of the more complex objects using more simple objects. The higher the level of hierarchy, the higher the gain in design time and also the higher the ease of data exchange with the other tasks of the system.

* The lowest level corresponds to the purely geometrical level of information, a description with basic features (line, arcs)
* A second level contains technological features, containing besides the geometrical information also technological information, i.e. the allowed methods for producing the specific component (rectangle, key_slot, obround, ...)
* The highest level contains functional features, describing a single or a group of components. All geometrical information can be derived from a limited number of parameters. A distinction can be made between:
    * A family (=parameterized) can only contain simple objects
    * A group (non parameterized) can contain any object of any feature type, apart from basic features.

The definition of each feature is stored in the fixed part of the database, this includes not only the geometrical but also the manufacturing information.

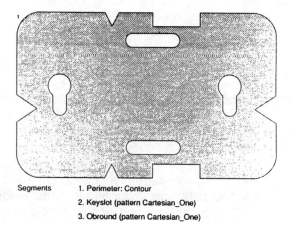

Segments
1. Perimeter: Contour
2. Keyslot (pattern Cartesian_One)
3. Obround (pattern Cartesian_One)

Figure 3: Example of the use of feature construct

5. DESCRIPTION OF KEY-MODULES IN THE SYSTEM (Figure 1)

The design module

The user is guided by a series of menus. Each menu is based on the definitions stored in the fixed part of the product database. Adding an element to the design consists in specifying the form feature to be used and assigning values to feature specific parameters, i.e. dimensions, positions, ... Simultaneously the workpiece description is stored in the variable part of the product database. The use of other functions like deleting or editing an existing form feature, also implies an automatic update of the workpiece description in the variable part of the product database, accessible from the other modules.

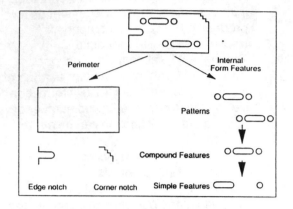

Figure 4: Feature representation of a workpiece

Initial tool selection

After design, an initial tool selection is made to get a first idea of the needed resources. This initial tool selection consists in retrieving the required tools for every method defined for each feature in the product database. The technique of parametric design with form features allows to relate the geometrical forms of the elements to the geometry of the tools by means of a set of rules. For each element of the workpiece, feasible tools are searched in the tool database, according to the methods defined in the common technical database.

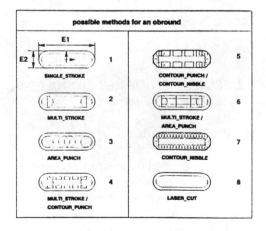

Figure 5: possible methods to manufacture an obround

Each manufacturing method is related to a set of rules. These rules enable the system

to decide whether a tool of a given shape can be used to produce a feature or a part of it.

In the figure below, tool-criteria are showed to manufacture an obround-feature according to a combined method: nibbling with a circular tool for the arcs and punching with a rectangular tool for the lines. For each segment of the feature, a relation is defined between the tool-parameters [T-values] and the feature-parameters [E-values] e.g. for the first arc, the diameter of the circular tool must be smaller then half the width of the obround-feature.

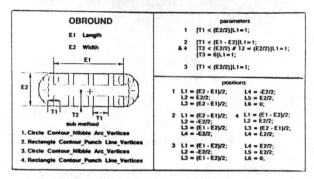

Figure 6: Production data of a feature

For all possible methods and tools, the punching coordinates, related to the origin of the workpiece are generated. The results of the initial tool selection are stored in the database.

Nesting of workpieces

The purpose of the nesting module is to combine rectangular workpieces on blanks with the aim to minimize material waste. In a first step, pieces are interactively clustered in rectangular enclosures. The most important feature of the clustering algorithm is the technique for clustering together 2 shapes to obtain a minimal area rectangular enclosure (prenesting). In the second step, the rectangular enclosures -called sets- are automatically nested on blanks. For solving the rectangular stock problem, the A.Albano and R.Orsinin heuristic method was chosen. The nesting-module keeps track of the availability of the blanks, calculates the material waste for each blank and the number of identical blanks to manufacture. The user finally determines if a nesting is to be accepted or not.

Final tool selection and technologic module

From the results of the initial tool selection module and the actual situation on the shop floor, a final tool selection is made for a specified nested blank. The goal of this module is to generate a proposal for the turret of a specific machine and to select the manufacturing processes to produce the actual sheet.
The rules of the final tool selection can be devided in two categories:

1) rules for the tool selection
   1. use tools from the standard turret
   2. use tools from the machine's actual turret
   3. minimize turret size
   4. prefer big tools

2) rules for the process selection
   1. select the method with the highest average priority
   2. prefer methods with a small number of different manufacturing processes
   3. prefer methods with a small number of different tools

Once the final tool selection is finished, the final punch locations are determined. A simulation of the tool path gives the user the ability to control the generated the tool path. The module has also specific options to add special technology data to the blank, like bridges, loops, read_in_lines, clamps ... .
The result is a cutter location data file, generated adapted to the machine controller.

generation of NC-code

To generate NC-tapes for different machines with different properties and different control, a generalised postprocessor has been developed. The postprocessor can be configured using configuration files and reads additional descriptions of the machines from the common database, e.g. position of clamps.

scheduling

The basic concept in this domain is the network concept. A production order is referring to a routing, which contains records for all activities needed to produce the item. An activity might be a normal manufacturing operation like punching, an assembly operation like welding or more soft activities like quality assurance, customer approval or engineering.
The production orders are inserted into the schedule-module. This module is a standard PC based package called PM-sim and generates, by means of "Gantt-charts", starting and ending times while selecting actual physical resources (machine/man) to perform the operations. The overall objectives of PM-sim is to balance out the conflicting objectives of:
* keeping delivery times
* minimizing throughput time for orders
* maximizing the capacity utilization

6. MANUFACTURING SYSTEM ON PC

For several years the advantages of Computer Integrated Manufacturing (CIM) have been put forward. Still, only a small percentage of the implementations really meet the expectations. In several cases a CIM implementation remains incomplete. It continues to consume available resources and manpower. Because of the large investments needed, only large sized companies can afford to enter the CIM-adventure. These problems were also WTCM's experience while integrating the complete integrated Sheet Metal Manufacturing system in industry. Since the key modules of the system required high software and hardware demands, the use of the system seemed not suitable for the smaller and medium sized companies. Even though the existing modules can fulfil the specific needs of a small company, some improvements are needed:

1) Limited degree of automation
   The larger CIM systems tend to overautomate, disregarding the operator. The control over the system should be remain at his side. The operator has in fact to make the final decision, supported by the information in the system.
2) ease of adaptation
   The system must be easily adaptable to the specific needs of each company. It must cope with new types of products and technologies. New machines and processes must quickly be included in the system

3) user-friendliness
   Small companies mostly have only limited resources. They can't invest large sums in training users or maintenance personnel.

In this context, WTCM is adapting the above described system to run on a low cost computer (PC) that is easy to maintain and to operate. Due to the seemingly high computational cost of full-fledged decision making algorithms, the system should now be viewed as a decision supporting system, rather than an automatic decision taker. On the other hand, the new system also allows a high degree of interaction in the decision process: at any moment, the user is able to impose his decision on the system. The main task of the system has been reduced to proposing and evaluating the possible alternatives and to carry out the job, once all decisions have been taken.

References

[1] Adamowicz M., Albano A., Nesting two-dimesional shapes in rectangular modules, Computer-Aided Design, Vol.8, No.2 April 1976, 27-33
[2] van 't Erve A.H., Generative Computer Aided Process Planning for part manufacturing, T.U.Twenthe Phd.Thesis 1988
[3] Integration of CAD/CAM and Production Control for Sheet Metal Components Manufacturing. BRITE p-2406-Reports
[4] Peters J., Van Campenhoudt D., Manufacturing Oriented and Functional Design, Annals of the CIRP Vol. 37/1/1988, 153-156
[5] Nordloh H., Knackfuss P., Thiel C., Hirch B., Integration of CAD/CAM and production control in sheet metal manufacturing - The basis for automatic generation of production data, IFAC-INCOM'92

1   BRITE project 2406 'Integration of CAD/CAM and Production Controle for Sheet Metal Components Manufacturing. The project ran from April 1987 until November 1991. The project team consists of research institutions from Belgium (WTCM) and Germany (BIBA), a Danish software and consulting house (Peter Mathiessen) and sheet metal manufacturers from Belgium (ETAP, Actif Industries), Denmark (Dronningborg Maskinfabrik) and Germany (BICC-Vero Electronic, Schichau Unterweser AG). A prototype of the complete system is installed at the site of Actif Industries, BIBA and WTCM.

INTRODUCTION OF COMPUTER AIDED PRODUCTION METHODS IN A COMPANY FOR AIR-CONDITIONING HANDLING UNITS

Ch. Kaup
HOWATHRM Klimatechnik GmbH, Brücken, FRG

P. Otto
Department of Informatics and Automation
Technical University of Ilmenau, Ilmenau, FRG

Abstarct. This article deals with the experiences in introducing CIM-components on example of an air-conditioning company. It is shown how the competition ability of small and medium sized companies can be improved by usage of appropriate information- and communication systems. The score of a priority oriented CIM-concept is formed by the product related individual software CAESAR, which serves for computer aided design of central air handling units and administers all technical, management and organizational datas. Besides, CAESAR also covers interfaces to CAD-, CAM-, CAQ-, and PPS-systems. All software systems are operated with 20 personal computers in a PC-Network. Central data base is the fileserver. The topology of the network is based on a star network. As transmission protocol the Token-bus-method is being used. The effectiveness of the gradually implemented CIM-concept has been proved based on various management index numbers.

Keywords. Computer aided technologies, CIM-concepts, small sized companies, computer networks, software design, air-conditioning

INTRODUCTION

After development and conversion of CIM-strategy in larger companies in the seventies and eighties (e.g. Bergmann et al., 1989, Kopacek et al., 1989) increased efforts were made to introduce computer aided organizational and productional methods also for small and medium sized companies. The increasing competition in the middle class forces the use of all possibilities of a reduction in cost. Besides the product quality mainly criterias as deliver reliability, contractual order handling and flexibility in adjustment to customer wishes form the preconditions for a succesful existance of small and medium sized companies. Reduction in cost in this fields are therefore unavoidable besides product related measures like rationalization, new products and specialization.

The most important possibility to this is to be seen in a mostly general inclusion of computer systems in all organizational and productional areas, whereby due to the minor possibilities of risk compensation a gradual initation should be looked for. After an analyses of strategies for computer integrated production in small and medium sized companies in the

following the successful conversion of a CIM-concept in the HOWATHERM Klimatechnik GmbH shall be demonstrated.

CONCEPTS FOR COMPUTERINTEGRATED MANUFACTURING IN SMALL AND MEDIUM SIZED COMPANIES

Under the term CIM all methods are combined, which enable a complete computer aided transaction of all management and technical missions of a company. It is subdivided in technological oriented functions as

- computer aided design (CAD)
- computer aided planning (CAP)
- computer aided manufacturing (CAM)
- computer aided quality insurance (CAQ)

and management oriented refering to quantities, target-dates and capacities as

- production planning and control (PPS)

Connecting element there is a continuos information flow, which considers the complete development process of a product as a unit. The introduction of CIM-systems in one step in small and medium sized companies is creating great difficulties in the hardware as well as in the software field because of various reasons (lack of capital, lack of experience and special knowledge). For small and medium sized companies it therefore seems to make sense to gradually introduce CIM within the scope of a hierarchical constructed modular concept. Therefor however a modualwise constructed software is necessary, which possibly should consist out of standard components in order to minimize training and maintenance efforts and to be able to fall back on already existing user experience (Lang 1990).

Today the linkage of independent C..-components is no longer unsolvable problem out of the technical point of view (Scheer 1986). The connection to a generally accessible database system is therfore not absolutely necessary for small and medium sized companies, whereby an immense accomodation effort of the individual components can be saved. The data integration should only be centralized so far as to make all necessary data completely and correctly available at the time of their processing at the working places, but the communication volume of the subsystems, one with another, is possibly minimal. Such a solution naturally is not available as finished product, but always has to be adjusted especially to the product pallet, the organisation structure and other criterias of the appropriate company.

It is therefore necessary that large parts of the software especially the interface software have to be prepared expressly for the special problem in this specific company. On the other hand it brings the advantage that with it the software and thus the whole CIM-concept can be adjusted to special requirements of the specific company and in a shortest period of time an economical and operative CIM-solution develops. With that at the same time the possibility of a gradually introduction starting with island solution is given. The enterprise specific user-software is the core of the CIM-solution and takes over the mission of a "coupling modul" (Gröner, Roth, 1987). At the same time it should take over the interface handling to the further necessary C..-components, which leads to a "priority oriented CIM-concept" (Kaup 1992). The priority of the branch and enterprise specific solutions can be on the market alignment of the company (Fig. 1).

The priority oriented CIM-concept has been transferred for the HOWATHERM Klimatechnik GmbH by development of productionrelated individual software CAESAR (Computer Aided Engineering Software for Airconditioning-Unit Research) and gradually converted into the daily routine (Kaup 1991, 1992).

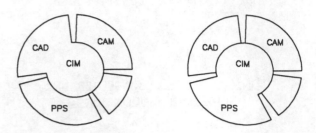

Fig. 1: Priority oriented CIM-concept

The concept of the priority oriented CIM-solution is based on subdivided data bases. This way each C..-component has its own data base, which is seen by itself selfsupporting. In such a concept of function integrated subsystems there are only few data bases in order to keep the communication effort between the subsystem possibly small.

In the field of the CA..-systems (technical and geometrical systems) the objected oriented file system is realizable, because non-standard data base systems are still in a research stage. In the field of PPS (organizational and management systems) standard data base systems are to be used.

Core piece of the solution is the product related C..-component, which represents the center of the process. From there the in the process chain integrated partial functions are called over shells as a child process (Fig. 2).

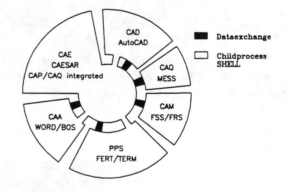

Fig. 2: Priority oriented CIM-concept in a medium sized company

SAMPLE OF THE USE OF THE PRIORITY ORIENTED CIM-CONCEPT

The introduction of CIM-components in small and medium sized companies can be illustrated at the example of a company for air-conditioning technology with 80 employees. In this company customer requirements are already considered at the time of construction, i. e. construction processes are incorporated in the order acquisition to make technical drawings available to the customer as early as possible and to examine the ability of realization of the requested specifications. Herewith the CAE/CAD field gains in a significance, also for the marketing oriented functions.

The manufacturing process of the company is complex and strongly oriented at customer requirements. The information expences in the manufacturing field are especially high. The integration throught of the informationsystems has here an especially high value.

In this case the strategy of the CIM-concept can be looked at in procedural understanding, it is therefor predominantly process oriented and to be automated strongly. Functions of the technical office are therby essentially integrated within the scope of order execution (Kaup 1987).

The information- and communication systems to be used within the scope of a linkage between office and manufacturing serves mainly for reducing the economical and technical risk in preparing offers, in reaching a higher exchange rate of offers into oders, in accelerating the construction action and in supporting the engineering field. A special aspect has to be considered in this case, the adjustment of the system in direction marketing in need of required variant manufacturing.

It is herbly easily overlooked, that today on the predominant consumer markets constructions and manufacturing can only realize what the field service brings along from the customer. The actual center of decision may therfore be at the connection point field service-customer (Mertens and Steppan, 1991).

In the model company very good preconditions for the integration of the individual systems were existing hardware wise, because all software-systems are operated in a PC-Network with then 13 today 20 personal computers. With a local network it is ensured that each user refers to the same data base.

On the hardware side the good starting position (LAN) should be kept. The complete conception should be limited to one computer architecture (micro computer / PC / LAN). As operating system the MS-DOS system is used uniformly, in order to be able to use every software system available at all computers.

The necessary accomplishment at the working place was given by adjusting the efficiency of the individual personal computers. In the CAD field exist much more efficient computers than for example in the text processing. Other computer architectures operating systems are to be refused because of the uniformity of the CIM-solution.

Basis of the CIM-concept of the model company is the special CAE-software. This software is the core

of the priority oriented CIM-conception and is outlined as a production related individual software. This software serves in the first place for the computer supporting layout of the air conditioning units and contains all product related, management and organisational datas. Besides this main function the software also covers the interfaces to further systems (Kaup 1991, 1992).

The goal is the total layout of the airconditioning units with the following results: sketch of the units, entire technical data sheet, detailed single data sheets of components, acoustic entire layout and appliance characteristic.

As further tasks follows the integrated management evaluation. Besides the calculated datas of the layout, especially the economical calculation of the unit, has to be considered for the customer, which compares in the background investment expenditures and operating costs of the system.

On of the integration modules of the CIM-management system establishes the connection with the PPS (Fig. 3).

The turnover of manufacturing list, e. g. list of parts, list of spare parts and generating of typeplates, the turnover of lists of parts for material economy, acquirment of order lists of parts to be baught, the turnover of standard time calculations to production planning and progressing pretain to that.

The moduls CAE-CAA encloses the takeover of commission texts to the invoicing as well as the acquirements of text datas and the turnover to text processing programms (Fig. 4).

The generating of CAD-compatible design datas and the turnover to the CAD-system are functions, which are realized by the modul CAE-CAD (Fig. 5).

CAE-CAP takes over the acquirements of cutting list and appliance specifications and sets up productions plans for the individual components. Controlling and optimization of cutting systems for profiles and plate bars is the task of the module CAE-CAM (Fig. 6).

Offer processing and order processing are comprised to project working places. In the regards to the organizational necessities it seemed to make sence, to desist from the existing organisation. Functional divisions of labor were able to be abolished. The goal was the creation of internal and external project administrators, were a project is accompanied to delivery of the order. Information losses between offer preparation and order processing are dropped.

The modified Interface DXF was use to connect CAE with the CAD software AutoCAD. The first three sections (Header, Tables and Blocks) are eliminated to reduce the size of the objects in the geometric database. In the field of the CA..-applications the object oriented filesystem, based on a standard ASCII-format, was used to guarantee a minimum

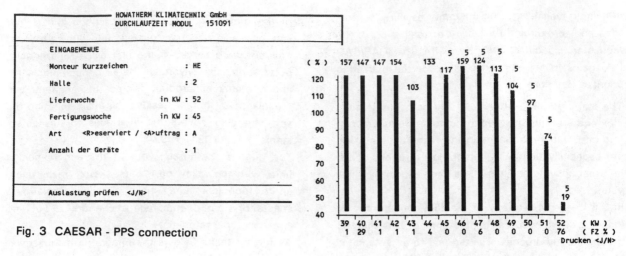

Fig. 3 CAESAR - PPS connection

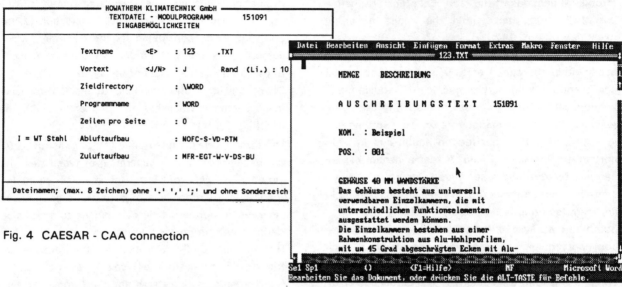

Fig. 4 CAESAR - CAA connection

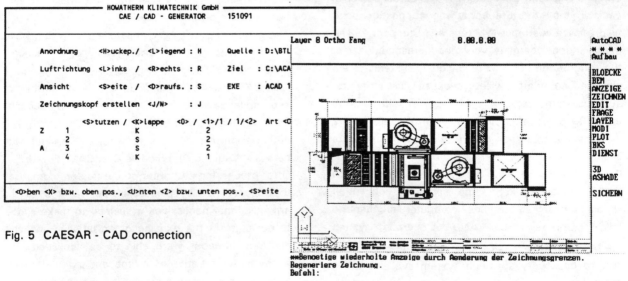

Fig. 5 CAESAR - CAD connection

standard of compatibility. Only in the field of PPS (organizational and management systems) standard database systems (ISAM) are used.

The integration level was not based on a unique database, because the requirements of the several systems are to different. It was based on specified transactionfiles to carry offer the information from one system to another.

The whole programm is based on a modular design. Each modul represents a real object, like heatexchanger (HWT), fan and motor (HV) or humidifier (HWA). In every modul all object datas (technical, organizational and administrational datas) are generated. All moduls are designed in Microsoft Basic 7.1 (Profi development system).

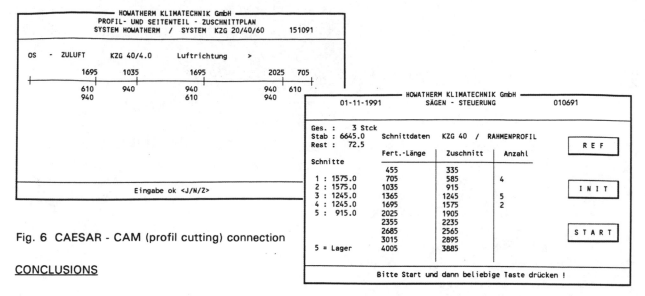

Fig. 6 CAESAR - CAM (profil cutting) connection

CONCLUSIONS

The priority oriented CIM-concept was realized gradually (3 phases) and implemented in large extent in HOWATHERM Klimatechnik GmbH Brücken. Core component is the enterprise specific, individually prepared CAE-application CAESAR, which is also taking over interface-handling to the offer- and order handling up to the manufacturing of air-conditioning central units. It was the goal to have mostly integrated task complexes handled by one projector.

The organizational separation between offer handling and order handling could be dropped. Thru the integration and extensive automation the total efficiency of the company increased in 1991 by 7.9 % (Son and Park 1987).

REFERENCES

Bergmann, S.; D. Mandler; J. Wernstedt; P. Otto (1989). Process Control system type PROFIS - a component of a CIM conception for fabricating microelectronic circuits. IFAC/ IFORS/ IMACS Symposium; Large scale systems: Theory and Applications, Berlin , Preprints, Vol. 3, pp. 420-424.

Gröner, L.; L. Roth (1987). CIM-Handler für die Verbindung von Softwaresystemen, in: CIM-Management, Ausgabe 3/87, München.

Kaup, C. (1992). Strategien zur Einführung der rechnerintegrierten Fertigung bei auftragsbezogener Einzelfertigung in mittelständischen Unternehmen, Konzeption und Realisation. Dissertation A, Ingenieurhochschule Mittweida.

Kaup, C. (1992). Der Weg zu CIM - Vorbereiten und schrittweises Einführen von Prinzipien der rechnerintegrierten Fertigung bei kleinen und mittleren Unternehmen. Maschinenmarkt, 14, S. 23-29, Vogel Verlag und Druck KG, Würzburg.

Kaup, C. (1992). Eingeführt - Realisieren der rechnerintegrierten Fertigung in einem mittleren Unternehmen.. Maschinenmarkt, 18, S. 88-91, Vogel Verlag und Druck KG, Würzburg.

Kaup, C. (1991). Entwicklung einer CAE-Software als Grundlage eines CIM-Managementsystems für den Mittelstand unter besonderer Berücksichtigung der Kopplung geometrischer und nicht-geometrischer Datenbanken. Diplomarbeit, Ingenieurhochschule Mittweida.

Kaup, C. (1987). Spezifische Vertriebsprobleme im ereich der industriellen Lüftungs- und Klimatechnik am Beispiel der Firma HOWATHERM Klimatechnik GmbH, Diplomarbeit, FH Ludwigshafen.

Kopacek, P.; P. Otto; J. Wernstedt (1989). A consulting system for the computer-aided operative production planning and manufacturing control, Preprints IFAC/IFIP/IMACS Symposium on Skill Based Automated Production, Vienna.

Lang, G. (1990). Auswahl von Standardapplikations-Software, Organisation und Instrumentarien, Berlin, Heidelberg, New York, London, Paris, Tokio, Honkong.

Mertens, P. and G. Steppan (1991). Die Ausdehnung des CIM-Gedankens in den Vertrieb, in: CIM-Expertenwissen für die Praxis. München, Wien.

Scheer, A.-W. (1986). CIM, der computergesteuerte Industriebetrieb, Berlin, Heidelberg, New York, London, Paris, S. 186 ff.

Son, K. and C.S. Park (1987). Economic Measure of Productivity, Quality and Flexibility in Advanced Manufacturing Systems, in: Journal of Manufacturing Systems, 3/1987, S. 193-207.

COMPUTER AIDED PLANNING FOR SIMULTANEOUS PART SHEDULING IN A FLEXIBLE MANUFACTURING CELL

Th. BORANGIU, A. HOSSU, A. CROICU

Department of Control and Computers, Polytechnical Institute
of Bucharest, Romania

Abstract. Computer Aided Manufacturing in flexible manufacturing cell (FMC) requires to consider in production planning problems related to part scheduling and tool transportation. The paper presents a method for solving the production planning problem in FMC, taking into account the following features: the capacity to select a certain resource (machine tool, MT) from several which can perform an operation, in each machining stage; the capacity to schedule the parts flow between MTs. In the paper a general manufacturing structure is considered, which consists of M multifunctional MTs and a number N of parts which must be processed in the FMC. The optimal planning is formulated as follows: find the sequences of the part feeding and machining such that the overall machining time for a lot of parts is minimized. A decision policy is defined as a strategy for selection of a particular MT from a number of available alternatives. The paper offers both an algorithm for part routing according to an overall performance criteria, and a research tool for an application setup and simulation. A set of conclusions based on experimental results are also included.

Keywords. Machine tools; Scheduling problem; Suboptimal solution; heuristic methods; human interactivity.

INTRODUCTION

In general, the problem of finding the optimal planning is not trivial from a computational point of view.

In order to solve this problem have been developped several techniques like: "exact methods" (or general purpose methods), exhaustive search, random search and heuristic methods.

Regarding to the other techniques the heuristic methods offers the capability to deal with "large scale problems" (the number of MTs and parts are very large).

Typically, the approach of this method is to sacrifice optimality for the sake of the speed, and the algorithm proposed in this paper belongs to the class of procedures which use heuristics in order to reduce the search effort.

The problem of production planning, usual is been resolved taking into account a set of assumptions (these assumptions are refered to setup time values of MTs for performing a specified operation):

(i) setup time values of MTs are ignored;

(ii) setup time values of MTs are considered not dependent of the sequences of part feeding and machinig. This hypothesis reduces the production planning problem to the (i) assumption by adding (for all the operations) the time required for setup the MT to the machinig time;

(iii) setup time values are dependent of the sequences of part feeding and machining but they are not separable of the part associated to the specified machinig operation. In this case the MT setup operation can not be iniatiated while the certain part is not ready for machining on this MT;

(iv) setup time values are dependent of the sequences of part feeding and machining and this values are separable of the part associated to the specified machining operation.

The algorithm presented in the paper represents a heuristic method for finding the sequences of part feeding and machinig, taking into account the last hypothesis (iv).

The system is provided with a graphic interface. This graphic interface offers to the human expert a set of tools in order to interact with the state of the algorithm and to act upon its evolution.

The method presented proposes to use the advantage of the interactivity between a production planning algorithm and a human expert in order to resolve this problem of production planning.

THE CONTEXT OF THE PROPOSED ALGORITHM

The initial data for this algorithm of the optimal planning problem solving:
- multifunctional machine tools MTs;
- parts to be processed.

The parts type data (the data belonging to the parts) are:
- number of jobs for each part;
- sequence of jobs (the set of MTs which can perform each certain operation);
- processing time values for each type of operation on MT;
- setup time values for each MT and for each job and part taking into account the job and part previous processed.

The machine tool type data represents the number (M) of the processing workstations included in FMC.

In developping the algorithm we make the following assumptions:

(i) All of jobs are not preemtive.

(ii) Values of setup times and processing times are a priori known.

(iii) The setup times of MTs are separable of the processing time, that means that it is possible to initiate a setup operation on the MT which can perform the next job of the part (if this MT is "free", i.e. MT is in the stage of waiting a job). This assumption give the scheduler system the possibility to initiate a setup operation even if the previous job of the certain part has not being finished yet.

(iv) Any MT can perform only a single activity (processing part or setup operation) at one time.

(v) The setup operation of the MT must precede the part processing operation without any other activity inserted between them.

(vi) The sequences of jobs (the set of MTs which can perform each certain opera-

tion) is a priori known. Is taken in consideration the possibility to perform multiple processing operations of the same part on the same MT

(vii) The possibility of a MT failure is not taking into account.

(viii) Any of the parts can be accessed in any stage of the FMC system.

(ix) In FMC can live several MTs which are able to perform the same job of a part.

The algorithm looks for minimizing the total time value (Tmax) required to process all of the N parts by a number M of MTs in the FMC.

Tmax if defined as being the maximum work time ($t(k)$) af all the MTs from FMC. The work time of a MT is here defined as the amount of the setup times, processing times and the waiting times (in the stage of waiting a job), times accumulated on this machine tool for a certain sequence of part machining.

$$Tmax = max \{t(k)\}, \quad k = 1,2,...,M$$

The algorithm has two stage:

- the internal stage - at the end of this stage the algorithm provides its own sub-optimal solution for planning problem;

- the interactive stage - in this stage the human expert has the possibility to modify the initial solution provided by the internal stage of the algorithm. Any action of the human expert is evaluated by the system (the modification performed upon the initial solution is tested if is valid or not);

Any modification upon the initial solution can be considered by the system as initial conditions and the first stage of the algorithm can be restart and it will provide a new scheduling solution based on the modifications performed by the human expert.

DESCRIPTION OF THE PLANNING ALGORITHM

In the first partition of the algorithm to each MT are assigned parts to be processed. The decisions of the assigning a job to a MT is taken every time a machine tool becomes "free" (MT is in the stage of waiting a job).

From the set of parts which are waiting for processing, only one part is assigned to the MT "free" and this decision in done under the object of minimizing the Tmax.

The algorithm proposed utilize the following initial data and working variables:

Entry Data:

N: number of parts which must be processed
M: number of multifunctional MTs;

M(ij): vector (list) of the MTs which can performe the job j on the part i; M(ij) is a list of MTs and not a single element because the MTs are multifunctional;

S(ijpq): setup time value for performing the job q of the part p, in the case this job is preceded by the operation j of the part i;

t(ij): processing time of the job j of the part i;

S(OOij): setup time value for performing the job j of the part i, in the case this job is not preceded by other job on the MT (this is the first operation on the MT).

Working variables:

t   : current time value;

t(k): the moment of time when the MT with number k will become "free", or in the case that MT is waiting for a job, the value of time when the MT has been declared as "free",
$$k = 1,2,...,M;$$

E(k): the number of the part which is in processing operation on MT with number k,
$$k = 1,2,...,M;$$

W(k): vector (list) of parts, which represents the "Waiting Queue" for the parts which will be processed on the MT with number k,
$$k = 1,2,...,M;$$

N(k): vector (list) of parts which are in processing operation on one of the MTs existing in FMC ("Next Queue"), the next operation on this part being assigned to the MT with number k,
$$k = 1,2,...,M;$$

J(i): the number of the job in processing stage for the part with number i,
$$i = 1,2,...,N;$$

F: the set of MTs which are in waiting for a job state ("free MTs");

C: the set of parts which are in the "Waiting Queue" and represent the candidates for the next job to be included in the planning schedule. To declare a not valid candidate we force his time value to be the MaxValue.

THE SKELETON OF THE PLANNING ALGORITHM

STEP 0. Variables initalization
```
t = 0
t(k) = 0,  k = 1..M
E(k) = 0,  k = 1..M
W(k) = {}, k = 1..M
N(k) = {}, k = 1..M
for i  = 1..N do
   W(k1) = W(k1) + { i }, for k1 in M(i,1)
   J(i) = 0
endfor
```

STEP 1. Start the processing operations for the first jobs on MT
```
for k = 1..M do
   if W(k) <> {} then
      if card { W(k) } > 1 then
        S(h1,r1) = min { S(0,0,i,1) + S(i,1,i1,1) },
                                                for i,i1 in W(k)
      else
           S(h1,r1) = MaxValue
      endif
      C = { i3 | i3 in W(k) and k in M(i3,2) }
      if C <> {} then
          S(h3,r3) = min { S(0,0,i3,1) + S(i3,1,i3,2) },
                                                for i3 in C
      else
          S(h3,r3) = MaxValue
      endif
      S(h,r) = min { S(h1,r1), S(h3,r3) }
      if S(h,r) = MaxValue then
         h = the single element from the set W(k)
      endif
      t(k) = S(0,0,h,1) + t(h,1)
      J(k) = 1
      E(k) = h
   endif
endfor
```

STEP 2. Update the "Waiting Queue", take the next job and update the "Next Queue"
```
W(k1) = W(k1) - { E(k) },         for k1 in M( E(k), J( E(k) ) )
N(k1) = N(k1) + { E(k) },         for k1 in M( E(k), J( E(k) ) + 1 )
```

STEP 3. Test of no more jobs to be processed
```
if W(k) = {}, for  k = 1..M then
   go to STEP 9
endif
```

STEP 4. Update the "Free MTs" times and the Free MTs list
```
t(k) = min { t(i) }, for i in { 1..M }
F = {}
```

STEP 5. If current time is smaller then the "free MT" moment then update the current time, take the new job and update the "Waiting Queue" and the "Next Queue"
```
if t < t(k) then
   t    = t(k)
   N(k1) = N(k1) - { E(k) },      for k1 in M( E(k), J( E(k) ) + 1 )
   W(k1) = W(k1) + { E(k) },      for k1 in M( E(k), J( E(k) ) + 1 )
endif
```

STEP 6. If there are no more jobs waiting for being processed then declare the MT "free" and update the "Free MTs" times
```
if W(k) = {} then
   F = F + { k }
   t(k) = min { t(i) }, for i in { 1..M } - F
   go to STEP 5
endif
```

STEP 7. Look for the part (h) and job of the part (r) to be candidate for the next job
```
if card { W(k) } <> 1 then
   S(h1,r1) = min { S( E(k), J( E(k) ), i, J(i) + 1 ) +
                 + S(i, J(i) + 1, i1, J(i1) + 1) },
                                                for i,i1 in W(k)
else
   S(h1,r1) = MaxValue
endif
if N(k) <> {} then
   S(h2,r2) = max { min { S( E(k), J(E(k)), i, J(i) + 1 ) +
                    + S(i, J(i) + 1, i2, J(i2) + 1) } ;
                 ; t(k2) - t(k) - t(i,J(i) + 1) },
```

```
                    for i in W(k), i2 in W(k), and k2 in M(i2, J(i2))
    else
      S(h2,r2) = MaxValue
    endif
    C = { i3 | i3 in W(k) and k in M(i3,J(i3) + 2) }
    if C <> {} then
      S(h3,h3) = min { S( E(k), J(E(k)), i3, J(i3) + 1) +
                     + S(i3, J(i3) + 1, i3, J(i3) + 2) },
                                                       for i3 in C
    else
      S(h3,h3) = MaxValue
    endif
```

STEP 8. Take the decision: what is next job and part to be included in the planning schedule

```
    S(h,r) = min { S(h1,r1); S(h2,r2); S(h3,r3) }
      if S(h,r) = MaxValue then
        h = the single element from the set W(k)
      endif
    t(k) = max { t(k) + S(E(k), J(E(k)), h, J(h) + 1); t } +
                                                 + t(h,J(h) + 1)
    J(h) = J(h) + 1
    E(k) = h
    go to STEP 2
```

STEP 9. The end of the internal stage of the algorithm

```
    Tmax = max  { t(k) }, for k in { 1..M }
```

STEP 10.
The second stage of the algorithm is now started. The system activates the graphic interface for interact with the human expert. The system displays the initial solution provided by the first stage of the algorithm.

STEP 11.
If the human operator does not want to make modifications on the current solution then STOP (the current sequence becomes the final solution), else, continue.

STEP 12.
a) The system asks the human operator to identify the part and the job of the part whose position he wants to modify, and also the new position of this part and job of part.
b) If the modification imposed by the human expert is not a consistent solution then the algorithm generates a planning consistent solution taking the position specified by the human operator as a unmovable position, else, continue.

STEP 13.
The total time value (Tmax') is computed for the new planning solution obtained.

STEP 14.
The decision of the planning sequence is taken by comparing the total time for the previous solution and the total time of the current solution:

$$Tmax = min \{Tmax,Tmax'\}.$$

STEP 15. Restart the second stage of the algorithm
go to STEP 10.

CONCLUSIONS AND EXPERIMENTAL RESULTS

The algorithm presented in this paper offers a good solution for the problem of planning, even if the human operator does not interact. For a large number of application examples (with a random set of entry data), the solution offered by this algorithm regarding to the solution provided by an "integer programming" method it seems to be with an average lost in optimality about 5 %.

This lost in optimality, regarding to the difficulties in computing time and amount of memory requirements imposed by the problem of automated planning may be considered as acceptable.

The presented algorithm may provide an optimal solution but it not guarantees the finding of this solution.

The algorithms of this type are very used in "large scale" problems (a large number of MTs, a large number of parts to be processed and a large number of jobs on the parts), because they provide consistent and good suboptimal solutions. In large scale system case an "exact method" is not possible to be implemented.

Another advantages of this type of algorithms are the posibilities offered by it to be segmented an run in small work sessions.

Naturally, for "small scale" system an algorithm which utilizes an "exact method" is preferable, because the solution offers by it is the optimal solution.

REFERENCES

Bellman, A., Esogbue, A. O. and Nadeshima, I. (1982). _Mathematical Aspects of Scheduling and Applications_. New York, Pergamon.

Borangiu, Th., Hossu, A., Croicu, A., and Nis, C. (1991). _Analiza si Sinteza Sistemelor de Fabricatie. Conducerea integrata cu calculator_. I.P.B. Press.

Kumar, P. R. (1990). _Real time scheduling of manufacturing systems_. Prep. of 11th IFAC, vol. 9, Talin, Estonia.

Zhou, C., and Egbelu, P. J. (1989). _Scheduling in a Manufacturing Shop with Sequence - Dependent Setups_. Robotics and CIM, vol. 5,no. 1, 1989.

CHALLENGING A CIM PILOT-BASED TRAINING CENTER

A.M.STANESCU, Th.BORANGIU, S.I.BRADATAN, R.PATRAS

**Polytechnic Institute of Bucharest, Computer Process Control Dept., CIM Division
313 Splaiul Independentei , 77206 BUCHAREST - ROMANIA**

Abstract. The paper is dealing with the problem of higher education multidisciplinary program within ComputerIntegrated Manufacturing Systems topics (CIM.S). Its main contributions are concerned with a particular educational environment design. Two principal issues are discussed: structuring the curriculum on the basis of Concepts-Tree Analysis (CTA) and developing the advanced management of educational resources.

Keywords: CIM, training, computer-aided instruction curriculum, educational management

1.INTRODUCTION

Computer Integrated Manufacturing (CIM) is "a marvelously vague concept that defies casy definition" wrote Charles Savage in his recent book (Savage 1991) enhancing the new managerial viewpoint about CIM problem. Is this the truth ?

Since J.Harrington Jr., realizing that the major manufacturing functions were "potential susceptible to computer conrol", has proposed the attractive term of CIM (Harrington 1973), two decades of intensive, well-funded research work focused the attention on this specific topics. Some important visionary program, as for example ICAM (1979) or ambitious regional programs as ESPRIT for Factory of Future, contributed to the successful basic researches. Contextually, many of the top universities from US, Japan and Europe are involved in cooperative work under progress. As a natural consequence, the special training program within CIM areas is representing a challenging topics for higher education of brilliant young scientists, engineers and managers.

It is not out of interest that these new centrers are aiming at both training and high technology transfer towards industry. Their main objectives are concerned with:

- **TRAINING:** multidisciplinary, continuous, open system educational environment;

- **RESEARCH:** both basic topics and applications developing;

- **PROMOTING:** demonstration, consulting, incubating small and medium sized enterprises;

Despite of the above-mentioned vagueness of CIM concept, the encouraging efforts to develop curriculum, pilot laboratories and new system of professional interrelationships were successful.

However, to develop an efficient CIM centre is a quite expensive task, operating advantageously for power and rich universities.

Keeping in mind, that CIM.E oriented enterprise is a prioritary goal of 1990' industrial challenge, we are faced with the nonhomogeneous diffusion process of basic knowledge and specific know how in this dynamic field of interest. A chance for the young generation from different contries could be lost if the training of the concurrent engineering, lean production and information technologies impact on manufacturing will be not included smoothly.

In the paper some issues are to be discussed: the multi- approaches formulation of CIM concept, generating the "force- lines" of well balanced curriculum, the management principle of CIM Training Center resources, the cost-effective use of CIM- pilot laboratories to achieve a gradual growing of training activities.

2. CONCEPTUAL FORMAL TRAINING WITHIN CIM SYSTEMS

Many attempts have been done to come closer to a more accurate formulation of CIM concept. There is the major interest to focus the theoretical investigation due to the deep correlation between developing a concept tree in the area of CIM and structuring an optimum training system of this multidisciplinary higher education environment.

The state of the art of formal concepts for CIM is concerned with various partial contributions, aiming to convert the term of CIM into the concept of CIM.

ESPRIT Consotium AMICE proposal (ESPRIT 1989) consists of two statements:

› **general statement** (refering to the socio-economical impact)

CIM is a strategical orientation of using computers to effectively integrate the manufacturing enterprises in order to improve their competitive position on price, quality and delivery time. To reduce the life cycle of their products as well as to increase the customization for their clients are also prioritary goals.

› **particular statement** (refering to the informational system architecture)

CIM.OSA defines an architecture for building a computer supported and integrated manufacturing enterprise.

One could notice that each proposal of a research team or network is oftenly enhancing a constrained viewpoint, based on their proper investigation results. It is natural for research work, but it is not always useful for educational purposes. Thus, such a focused approach generated a linear way of thinking. Teicholz and Orr (1987) wrote:

"Computer Integrated Manufacturing is the term used to describe the complete automation of the factory, with all processes functioning under computer control and only digital information tieing them together.

In CIM, the need for paper is eliminated and so also are most human jobs.
Why is CIM desirable ? Because it reduces the human components of manufacturing and thereby relieves the process of its most expensive and error-prone ingredient."

One can recognize a quite characteristic oppinion of an automatonist expert, thinking linearly at the evolutive concept from Aristotle auto-maton towards "paperless and peopleless" factory. Two remarks are to be done reviewing many other references:

1. The FoF and CIM are two closely related new-born-together concepts;

2. A unique partial approach is not compatible with multidisciplinary basic characteristic of CIM systems

We propose for your critical comments the following two statements (Stanescu 1992):

DEF1: FoF is an evolutive concept to desctibe the vortex-like thinking for people menthality change, aiming the new human involvement in discrete-part oriented, intelligent, lean production in which creativity & flexibility are assured by professionals.

DEF2: CIM is the strategical concept to achieve the global objectives for developing FoF

This concept is based on a multi-approach, performing a holistic insight, based on:

° SYSTEM APPROACH, including "3D" concurrent analysis

● **discrete events dynamic system:** modelling, system concepts for quantitative performance criteria, algorithms for system decomposition/aggregation of sector subsystems

● **information system:** architecture, enterprise data modelling, product data exchange, data management

● **business system:** responsiveness of manufacturing enterprises facing market changes; achieving optimum business performance criteria

° LEAN PRODUCTION APPROACH

● **concurrent engineering:** product and process concurrent design including quality assurance and product service design

● **material flow engineering:** receiving, storage, handling, shipping

● **flexible production automation:** flexible manufacturing cells, flexible assembly cells, components (robots, workcenters, NC axis, PLC, sensors a.s.o.)

° MANAGEMENT APPROACH

- material management, total quality, Just-In-Time

- finance, sales, marketing

- integrating enterprises through human networking

By using the above-mentioned multi-approach a CONCEPTS-TREE could be built, as the basis for curriculum developing of the CIM Training Center. Different coursewares will cover correlated concepts, some of them being overlapped. This step is representing the educational system top-down analysis. A partial sketch of the concepts-tree is shown in fig.1

3. LEARN ONE FROM ANOTHER

The challenging topics of FoF/ CIM require skills and knowledge found neither in a syntetic discipline nor a simple summing collection of well entitled coursewares. To face with the CIM Training goals and objectives an **educational system** has to be activated.

"A CIMS encompasses the total enterprises for the design and production of goods from raw materials and components, using computers and communication systems to coordinate and unify the flow and processing of materials and information" is a basic philosophy of this training program.

Obviously, each student, working individually, "**will not be expert in every area rellevant to manufacturing instead, they must have knowledge and skills in a particular area, coupled with a general knowledge of the entire manufacturing enterprise and an ability to work well with the others**" there is a top educational principle (abbreviated as **team-work** or **team approach**), which is well formulated in CIMS-Student Manual of Georgia Institute of Technology (1990-1991).

On the other hand, the "peopleless" factory based vision if CIM is a "dead-end" because it leaves people out of the equation. The people, bright young engineers, managers and scientists, are those who gives to an organization its **flexibility** and **creativity**.

"**The specialists work together as a team** sharing their knowledge and **insight** not only to solve the complex problem of advanced manufacturing, but to come up with need and innovative offering"

wrote Ch.Savage in his opening-window book (Savage 1990).

According to the fundamental principles:

- cross-functional team-approach

- human networking enterprise integration

the present paper is proposing for critical comments a potential management idea for CIM Center educational resources.

Let us build a Training Ring, compatible with a company management ring (Savage 1990), even for small and medium sized CIM- oriented enterprises. We propose a partition for this ring in seven educational segmant concerning: **SYSTEMS// ENGINEERING// MANUFACTURING// MATERIAL FLOW// FINANCE// MATERIAL RESOURCES// MANAGEMENT**

Each segment has various application area, as it is shown in fig.2. Each application area is based on several disciplines. In Appendix the SYSTEM segment is presented as a sample of Training Ring.

Keeping in mind that the CIM Training Center is used for post- graduate 3 years higher education, one could imagine a SPIRAL INVOLVEMENT, consisting of three training cycles

- **1st cycle:** Preparatory cycle, guided by the objective "FIND YOUR PERSONAL WAY". Crossing segment by segment from SYSTEM to MANAGEMENT (1st year) each student should appreciate himself which is the proper segment - compatible with his prerequisites, a priori skills a.s.o. In the same time he is gathering some general knowledge of the entire manufacturing enterprise.

- **2nd cycle:** Working inside a selected ring segment for one year, the student will grow up to be a specialist for a specific topics. The main goal is dealing with "FIND THE WORK MOTIVATION INSIDE A TEAM". The student mobility should concern with different segment areas.

- **3rd cycle:** Finally, to prepare his M.Sc. thesis, each student, belonging to a specific area team should face with cross- functional tasks, involving all the teams. The goal could sound in the following manner: "FIND THE EMPOWER OF HUMAN KNOWLEDGE NETWORKING". A CIM project for SME will be selected, developed and managed in terms of a cost-effective use methodology.

4. CIM PILOT BASED LABORATORY

CIM System laboratory should look like a complete mini-factory, integrating all the important functions that are present in real-life production facilities. The main components of educational platform (TEMPUS-JEP, 1991) are planned to be:

- **FMC** - Flexible Manufacturing Cell - where parts are machined, prepared and inspected

- **FAS** - Flexible Assembly System - for both mechanical and electronic package

- **AGVS** - Automated Guided Vehicle System - carrying the pallets from/to the production cells, the receiving area and the shipping area.

- **RA** - Receiving Area - where parts that have to be machined or assembled are brought in the system or temporary stored

- **SA** - Shipping Area - for the products that have been produced within the system

The second key of laboratory configuration is concerned with the following two basic softawre packages:

- A **Cell-oriented platform,** providing a complete environment to develop all cell control functions. It will play the role of an application development platform to build cell control, providing high-level object-oriented interfaces to operators, application software modules, communication networks and plant floor devices like PLCs, robots, vision systems a.s.o.

- A **Factory-oriented platform,** providing a multi-vendor base upon which one can build and integrate factory automation software, including three hierarchical levels: plant/area/cell. It has to bridge the gap between Plant level business management computers and Shop Floor Industrial Control.

As concerning the computing infrastructure, the laboratory should integrate multi-vendor computers (workstations, token ring LAN, PC computers a.s.o.) but the networking with University Main Frame is useful to link with European Academic Research Network (EARN-Digital).

5. PERSONALITY MODELLING SPACE

To achieve the challenging requirements requested by a Center of Excellence for CIM, a well-balanced higher education could be developed by the New Renaissance-like Phylosophy.

Due to the fact that the subject is too complex and important we should like to suggest a new approach only. Keeping in mind that a quality of experts or technocrats will be not satisfactory to prepare smoothly the 21th Century global civilization is to aim at rediscovering the harmony of human personality. To face with the danger of creating "unidimensional human being" (or, even worse, non-dimensional humans) as Marcuse (1977) said, the evolutive directions od a student-professor peer enhancement should imply:

- **High Tech Career**, CIM fields being an outstanding field of interest

- **Socio-Cultural Background**, based on lectures and conferences on History of Civilization, Sociology, Econometrics, Psyhology a.s.o.

- **Metaphysics Background** based on lectures and practise on Phylosophy, History of Religions, Music and Fine Arts, Poetics a.s.o.

6. CONCLUSIONS

The present paper is based on two-years intensive work of preparing the TEMPUS-JEP entitled "Formal Training for CIM Systems".

The key idea is to create a challenging educational enviroment for higher education of brilliant scientists, engineers and managers for the Factory of the Future. To succed in such an important task, the networking of Training and Technology Transfer Center within CIM from every important university metropolas seems to be an interesting solution.

7. ACKNOWLEDGEMENTS

In many respects the writing this paper is a result of a cooperative effort for underlining the necessity to promote CIM training program all over the world. We are particularly indebted to Prof. Wayne Book, Georgia Institute of Technology, who stimulated our imagination during direct discussions or by his papers and books. Also we are deeply grateful to Prof.Herbert Schulz, Director of the Institut fur Productionstechnik und Wergzeugmaschinen (PTW), Technisce Hochschule Darmstadt who act as Coordinator of our TEMPUS-JEP

Last, but not least, we are grateful to Ms. A. Rosu acting as Director of SIB for her provoking cooperative work (Stanescu, 1992)

8.REFERENCES

Savage, Ch.M.(1990). 5th Generation Management, Digital Press, Bedford

Harrington, J.jr.(1973). Computer Integrated Manufacturing, Industrial Press, New-York

(1979).Integrated Computer Aided Manufacturing, ICAM Program Prospectus", Wright Patterson Air Force Base

ESPRIT Consortium AMICE.(1989).Open System Architecture for CIM, Springer Verlag, Heidelberg

Teicholz E., Orr J.N. (1987). Computer Integrated Manufacturing Handbook, McGrow-Hill, New-York

Stanescu A.M., Bradatan S.I. (1992) Training requirements for CIM systems Career, Technical Report, Polytechnic Institute of Bucharest Press, Bucharest

(1990-1991) CIMS Student Manual, Georgia Institute of Technology, Atlanta

(1991) TEMPUS-JEP 4726 proposal: Formal Training for CIM System within Romanian Universities

Marcuse H. (1977) Phylosophycal writings", Politics Publ. House, Bucharest

Stanescu A.M., Borangiu Th., Rosu A., Bradatan S.I., Costin A. (1992). Organizing a Pilot-Center for SME/CIM.E training program - Technical Report granted by the Ministry of Labour, SIB Press

Appendix:

1. SYSTEMS

1.1. DEDS

(1) Discrete Events Dynamic Systems Analysis
(2) Models in System Engineering
(3) Simulation of Manufacturing Systems
(4) Digital Systems Design
(5) Analysis of Algorithms
(6) Queueing Theory/Location Theory
(7) Stochastic Process Control
(8) Feedback Control, Design and Implementation
(9) Applied Regresion and Time Series Analysis

1.2. INFORMATION TECHNOLOGY

(1) Information System Design
(2) Advanced Computer Organization
(3) Computer Networks / Local Area Networks
(4) Data Base Design
(5) Computer Communications System
(6) Computer Operating Systems
(7) Human Computer Interfaces
(8) Knowledge based Systems
(9) Object-oriented environment Design

1.3. BUSINESS

(1) Management Information System Methodology
(2) Modern Organisation of SMEs
(3) Advanced Simulation
(4) Analysis of Production Operations
(5) Management Application of Artificial Intelligence
(6) Manufacturing Planning and Control
(7) Analysing of marketing scenarios
(8) Quality Control in Manufacturing Systems
(9) Financial / Managerial Accounting

DIGITISING AND MEASURING OF FREE FORMED SURFACES

Barbara Pogorevc, Dipl. Ing.
Tehniška fakulteta Maribor, Laboratorij za tehnološke meritve
Smetanova 17, 62000 Maribor, Slovenija

Abststract. According to the modern CIM concept a tendecy exists
to data connections between different CAx systems.   Therefore
the coordinate measuring technologie of the future will be based
on data connections with the CAD/CAM systems. The programming of
measurements and digitising of complex parts will become simple.
Key Words. Coordinate measuring technologie; measuring;
digitising; free formed surfaces

1. Free formed surfaces

Workpieces, built of regular
geometrical elements (circle,
cylinder, cone, straight line
etc.) are mathematically
simple to describe. At
present the measuring of such
workpieces is a routine work.
Therefore the interest of
coordinate measuring
technique moved to the area
of free formed surfaces.

Workpieces with free formed
surfaces and regular formed
workpieces differ one from
the other not only in
geometrical characterisctics
but first of all in duration
of theit development,
optimisatiopn and production.

A lot of usable articles are
built of free formed
sufraces:
- stream exposed objects
- ergonomically designed
objects
- estheticaly designed
objects
- functionally designed
objects etc.

Free formed workpieces are
developed mostly by the
design or experimental work.
Because of the wide use of
such methods, a requirement
exists for qick, reliable and
inexpensive measuring and
digitising devices.

Characteristic for workpieces
which are developed by design
or experimental work is the
nonexistance of their
mathematical description. It
exists only their physical
model.

From this fact results the
basic measuring - technical
difference between regular
and free-formed surfaces:
- regular geometrical
surfaces: the mathematical
description is known,
therefore the form defines
measuring points
- free formed surfaces: the
mathematical description is
unknown, therefore measuring
points define the form.

Two possibilities are
disposed in cases, when
instead of the mathematical
description only the physical
model exists:

- direct production of new workpieces on copying milling machines or multiplication processes (casts)
- digitising.

The production is in the first case relatively simple regarding the measuring. The weakpoint of the milling are long lasting processes because the physical model must be totaly scanned for every new workpiece. Got workpieces are perfect copies of the physical model including all inexactnesses.

Workpieces which were made by milling or multiplication processes can be controlled only in a few measures or expensive etalon measuring instruments are used.

Therefore in the area of free formed surfaces the digitising prevails more and more as an actual method to define the mathematical description of the surface.

The main advantage of digitising before conventional methods is the exchange of the physical (hardware) model with the mathematical (software) model.

Digitising is a relatively long lasting process because of the great number og points which have to be perceived. Sufficient number of perceived points is the only assurance for the exactness of the mathematical model.

In opposition to the direct production methods digitising requires scanning only once for all workpieces.

A lot of programme systems were developed for the quick and simple reconstruction of mathematical models of free formed surfaces (HOLOS - Zeiss, SURFER - DEA etc.)

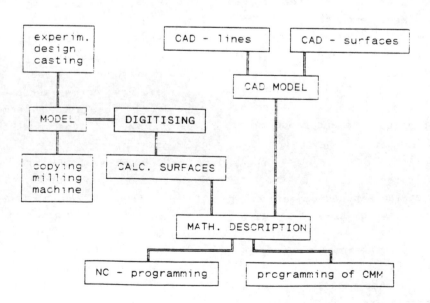

Fig. 1: Possibilities to production mathematicaly undescripted free formed workpieces

2. Measuring of free formed surfaces

The measuring is defined by the cintinuance of following steps:
- generation of coordinates of points
- calculation of deviation between real and ideal values
- numerical and/or graphical presentation of results.

Described cintinuance of steps should actually be named controlling, but instead of this the term measuring advanced.

Generation of points can be carried out by scanning curves or by scanning particular points. The position and the density of points depends on the given task, engineer's experience and his judgement.

A lot of programme systems enable collision control on the computer screen. This possibility is very important for measuring programming of complex workpieces with extensive probe combinations.

3. Digitising of free formed surfaces

Digitising is defined as the reconstruction of real geometry of the mathematically unknown workpieces. According to this definition this process should be named measuring, but this term prevailed for the control.

There are two ways to define the mathematical description of unknown surface:
- on-line connection between the digitising device and computer
- off-line connection between the digitising device and computer

The exactness of the mathematical description depends on the number and the position of perceived points. The measuring engineer defines these points according to hid experience only.

The on-line connection enables the interative process which is repeated until the difference between the real and ideal values is smaller then the prescribed tolerance.

4. Conclusion

A tendecy exists to the data transferring with standard data interfaces (VDAFS, DMIS, STEP etc.). Therefore a lot ofmeasuring and digitising devices are based on the common CIM concept and on the use of standard data interfaces. This characteristic enables data connections among different computer systems. Manual intervention in the data transfer becomse nearly superfluous.

5. Literature

H. Weule, H. Klein
Messen und Digitalisieren von Freiformflaechen unter Einsatz von CAD - Systemen, VDI Bericht 836, VDI Verlag Duesseldorf 1990

S. Bruno, W. Grebe
Freiformflaechen graphisch - interaktiv digitalisieren, Sonderteil in CAD-CAM-CIM, Mai 1990, C. Hanser Verlag Muenchen

H. D. Jacoby
Kopplung von Software fuer KMG mit CAD/CAM Systemen, VDI Bericht 540, VDI Verlag Duesseldorf 1984

DESIGNING OF NEW MEASURING EQUIPMENT WITH CAD-SYSTEM RESPECTIVELY

WITH FINITE ELEMENT METHOD

Miroslav ŠEROD, M.Sc.Mech.E. *

* Technical Faculty Maribor, Laboratory for Technological Measuring
  Systems. Smetanova 17, 62000 Maribor, Slovenia

Abstract. Designing of measuring device (Height measuring instrument)
with CAD-method: Finite element method (FEM) will shown in this paper.
FEM is able to define deformations and tensions of parts of measuring
devices and to find the optimal constructions.

Key words. Designing of measuring devices with CAD-methods; Finite
element method (FEM); Height measuring instrument (HMI); mesh of measuring parts; deformations and tensions; three-dimensional finite elements: EZ60 - rectangular prism and EZ45 - triangular prism.

1. INTRODUCTION

In the Laboratory for Technological Measurements at the Technical Faculty in Maribor
an automatic controlled measuring device
for two-coordinate height measurements -
Height Measuring Instrument (HMI) has been
developed and manufactured by a team of experts. The aim of the development was investigation of the accuracy of the measuring system. Tensions and deformations on
more important parts of measuring devices
were analysed with the aid of a computing
method, the so called Finite Element Method
(FEM). This method was used as a construction resource for the execution of such
constructional parts, which will assure as
precise measuring of dimensions, forms and
positions as possible.

2. RESEARCH OF STABILITY AND ACCURACY OF HMI ACCORDING TO THE FEM

The stability of the constructional parts
of HMI can be analysed by the known FEM.
With the aid of computational analysis,
stresses and strains of each single element
can be determined, showing their stability,
for it is almost impossible to calculate
the complex design elements using the
strength method. Some parts of the HMI are
rather complex, geometric shapes. The FEM
proved to be suitable for solving such
design problems.

Our measuring device (HMI) enables two-
coordinate measuring of parts. We have
found out that most measurements reach a
height of 750 mm. This is why the range of
measurements has been determined up to 750
mm. The electronic HMI was designed for
computer aided control of the measurement,
enabling us to monitor current operations
and measured values on a LCD (liquid chrys-
tal display). A measuring protocol can be
printed for all applied measuring operations and results can be evaluated.

Figure 1 shows HMI and all the influences
that had to be eliminated from the design
and program standpoint in order to obtain
the desired measuring accuracy.

The HMI will be used to measure dimensions,
i.e. heights, depths and distances as well
as deviations of forms and positions, like
straightness and squareness of the measured
part.

With the aid of FEM the following four main
constructional elements of electronic HMI
were calculated: pillar, socket, carriage
and the air bearing.

For calculation according to FEM, the following steps have to be taken:

- prepare geometric data on finite element
  mesh,
- define and determine, resp., matrix for
  stiffness of single elements,
- unite matrics for stiffness of single
  elements into the global matrix for
  stiffness of construction,
- solve the system of linear equations and
  calcute constriction displacement,
- indicate analytic and graphic results of
  calculations, which represent stresses
  and strains in the investigated construction.

During the investigation of HMI parts by
FEM, we found out that, regarding the
demanded accuracy of the whole HMI, it is
much more suitable to apply more demanding
analysis by FEM. It means that we used
models with an enhanced mesh of finite
elements; simultaneously more demanding
three-dimensional analysis including apli-
cation of space finite elements was used.
It is actually impossible with simple two-
dimensional analyses to include all influences which have an effect on different
points of the examined part. Regarding
previous experiences with FEM we decided to
apply three-dimensional finite elements,

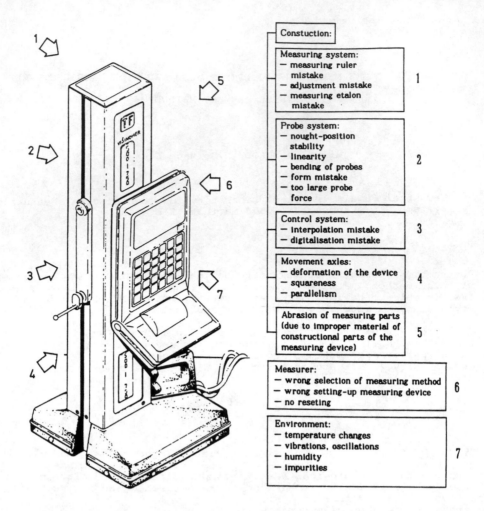

Fig.1.   Height measuring instrument (HMI) with a computer-unit and
presentation of all influences affecting it

type EZ60 and EZ45 with the aid of a model. The model, presenting a certain part of the HMI, had to be built. Both types of elements enable a reliable simulation of processes in busy parts of the HMI and at the same time optimal utilisation of accesible computer memory and time. With good planning of finite element mesh all critical and disputable spots of selected construction can be discovered, preventing the unpleasant consequences of an incorrect construction.

2.1. PILLAR OF HMI

The calculation by FEM assumes the pillar to be made of alloyed manganese-vanadium tool-steel with an elasticity modulus $E = 210000$ N/mm2, the refencial temperature of $20\degree$ C and all DIN prescriptions for permissible temperature corrections. In our calculation 84 space elements type EZ60 were applied with the aid of a pillar. See the pillar mesh in figure 2.

The duration of the calculation in the central processing unit of DEC-VAX-8800 was app. 15 minutes. Analysis of the results showed that the construction of the pillar was deformed, as had been expected. Due to the surface loads on the upper part, the pillar bent in the direction of active point loads at a max of 0.002 mm at the measuring force of 1 N for 0.00041 mm at the measuring force of 0.2 N. Shortening along the height was of no importance. Displacements were truly very small, however decisive for measuring instruments due to the demand that their accuracy class was to be within a thousandth of a milimeter +/- 1 μm.

2.2. SOCKET OF HMI

The most loaded of the fixed socket is the upper part where the pillar of HMI is installed. Pressure to the surface amounts to 0.113 N/mm2. Loading due to the grip which will be fixed to the socket and used for handling with the HMI (e.g. moving, pressing, shifting, etc.), has been considered as well.

The socket is made of cast iron with an elasticity modulus $E = 100000$ N/mm2. The ambient temperature in the measuring room and workshop was considered in our calculation. The socket mesh is shown in figure 3. For our calculation app. 300 space elements type EZ60 and EZ45, which are the constituent parts of the socket, were used.

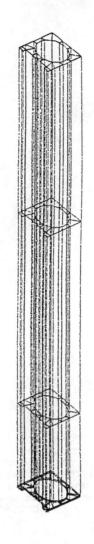

The duration of the calculation on the CPU of the computer VAX-8800 was app. 60 minutes, terminal-time was a little longer.

Analysis of the results showed the expected deformation of the socket, with forces of 140 N (weight of the pillar and barrow of HMI, including all elements) and of 300 N (grip handling), were calculated. On the upper part of the socket too large deformations and continuous spread displacements of 0.006 mm were found. On the upper part, on the spot where the pillar is installed, the socket was almost bent due to stressing. The decisive influence of stressing proved to be over grip. During the course of the second calculation the smaller stressing over grip 30 N was taken into consideration i.e. the usual one at the correct handling. We noticed that smaller displacements appeared on the same spots, the greatest amounting to 0.0015 mm. Nevertheless, they were still too large, regarding the permissible value of 0.001 mm. Due to this fact the initial design of the socket had to be changed. After thorough analysis we decided to add two reinforcing ribs next to the grip, see figure 4.

The FEM calculation showed considerable precedence of the new construction over the former one. The max. displacement under the loading of 300 N was app. 0.001 mm and deformations were distributed more continuously. The selected construction of the socket was entirely adequate for a prototype of HMI.

Fig.2. HMI pillar mesh for analysis by FEM

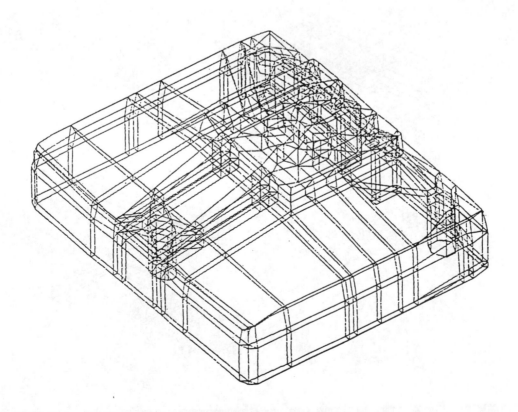

Fig.3. HMI socket mesh for analysis by FEM

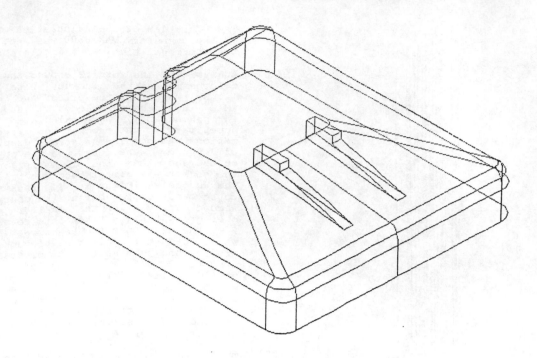

Fig.4.   HMI new socket with two reinforcing ribs

2.3. CARRIAGE OF HMI

The carriage mesh is shown in figure 5. The carriage runs along the pillar over bearings. An electric motor (5 W) drives a steel band on which the carriage is fixed. The band can move in two different directions, lifting and lowering the carriage to the selected height.

Due to tensile force the loading will be transfered over screws from the band to the wall of the hole on the lower carriage fitting. The max. load appears when the band moves up, i.e. when the carriage moves upwards. In this case not only the tensile force but also the weight of carriage with all its parts are acting on the wall of the hole (app. 15 N). Loadings during the lowe-

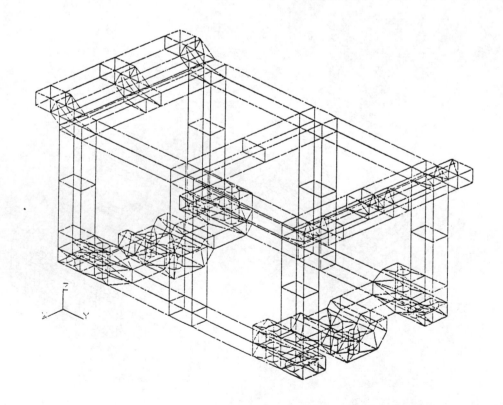

Fig.5.   HMI carriage mesh for analysis by FEM

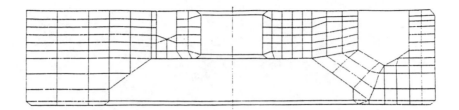

Fig.6.   HMI air bearing mesh for analysis by FEM

ring of the carriage are smaller because only tensile forces acting on the wall of the hole are considered and the weight is neglected. The total pressure to the surface amounts to 3.05 N/mm2. The first calculation by FEM consideres only surface loading and the second one surface loading plus two point loadings (150 N) which appear during carriage handling. The carriage is made of alluminium alloy with the elasticity modulus E = 70000 N/mm2. Ambient temperature was taken into consideration.

The analysis of the results FEM proved the carriage structure to be deformed as expected. The largest displacement of 0.0074 mm appeared in the carriage socket in the direction of band tensile force. Stresses can be neglected as loads are not large.

2.4. AIR BEARING OF HMI

The two-dimensional mesh of the air bearing is presented in the figure 6. HMI has three air bearings which are placed under the supports on the inner side of the socket.

The air bearings are fixed to the supports by screws. They support the whole weight of the HMI, each of them carrying 1/3 of the total weight (70 N). The calculated pressure of surface loading amounts to 0.26 N/mm2. For our information pressure at 200 N was calculated as well. The results was 0.732 N/mm2. While preparing input data, certain simplifications had to be made. Air bearing is a rotational solid which can be examined in cross-section. In our case, the most unfavourable cross-section was chosen, presenting the weakest part of the structure. Bearings were made of alloyed chrome-molybdenum steel with an elasticity modulus E = 210000 N/mm2. The calculation considered the circumstances at referential temperature and ambient temperature in workshop.

In our calculation 165 surface finite elements were used to build the structure of a bearing. Analysis of the results showed expected deformation of the construction. In the first calculation, with burdening of 70 N, the largest displacement was at the upper edge of the bearing by 0.00043 mm.

In the second one, with a loading of 200 N,

it appeared on the same spot with a displacement of 0.0012 mm, i.e. exceeding permissible deviation. The bearing, however, will not be loaded in such a way. Strains were mainly continuously distributed, while stresses were negligibly small. The chosen construction of the air bearing therefore proved to be quite adequate.

3. CONCLUSION

The prototype of the HMI is the basis for the development of coordinate measuring devices in Slovenia. Its advantage is that, provided the design solution is optimal, the manufacturing is technologically possible. Following CAD/CAM principles, the elements of domestic production were machined on a CNC machining centre. The calculated purchase price of 15000 DEM is competitive on the measuring equipment market, for with the HMI we are measuring in two-dimensional system, although its software is the same as the software of simple three-dimensional device.

4. REFERENCES

1. Šostar, A. & Co. (1988). Project of HMI-Height measuring instrument. Project task of the Labor for technological measuring systems on Technical faculty in Maribor.
2. Šostar, A. (1988). Technological measuring systems. Notes on lectures. Technical faculty Maribor.
3. Šerod, M. (1990). Influences on precision of measuring instruments on the example of an Height measuring instrument. Technical faculty Maribor.
4. Prelog, E. (1975). Finite element method. Faculty for mechanical engineering (FME) in Ljubljana.
5. Prelog, E. (1978). Elasto- and plasticity mechanics. FME in Ljubljana.
6. User's guide for FEM-BERSAFE program. (1982). Technical faculty Maribor.
7. Hellen, T.K. (1970). BERSAFE (phase 1) - A computer system for stress analysis. User's guide. Berkeley nuclear labors.
8. Sekulović, M. (1988). Finite element method. Faculty for mechanical engineering in Belgrade.

ALGORITHM AND SYSTEM FOR AUTOMATIC CAMSHAFT TESTING

T.C. IONESCU and S.M DASCALU

Polytechnical Institute of Bucharest, Faculty of Control and Computers, Department of Control and Industrial Informatics,
Spl. Independentei nr. 313, Bucharest 77206, ROMANIA

Abstract. An original method for testing camshafts is developed. It can be used for a large variety of engines and it has been implemented for several Romanian engine manufacturers. The testing algorithm is presented and a discussion concerning its merits is made. The paper also containsin an outline of the hardware system used in conjunction with the algorithm. It is worth mentioning that the practical results are more than encouraging from the point of view of accuracy and efficiency of testing.

Key Words. Camshaft; automatic inspection; zero acceleration zones; intelligent interface.

1. INTRODUCTION

Our research was directed towards developing a family of systems dedicated to the camshaft automatic inspection and testing (Dascalu, 1989). The starting point was characterised by a marked absence of information concerning technological aspects and, moreover, a perfect disagreement about the appropriate testing methods. We have used the general information obtained from the Italian company Samputensili Bologna (Anon., 1982) and the preliminary results offered by Pitesti University (Anon., 1983). A first set of conclusions has been drawn in an early paper (Ionescu, 1984), in which the importance of the proper camshaft testing for the operation of an engine is emphasised. It has been shown that the profile errors and the contour undulations of cams, resulting from inaccurate processing, induce vibration phenomena and loss of engine power.

The main objectives of the camshaft automatic inspection and testing aim at determining the geometric parameters associated to the camshaft. These parameters are:
- real positions of cam reference axes (expressed in angular values);
- relative positions of inlet and exhaust cams, i.e. the so called "angular offset";
- rising diagrams of the probe on the cams;
- speed diagrams;
- acceleration diagrams;
- bearing journals "excentricity" (position errors of journals centres with respect to the rotation axis), expressed in polar coordinates;
- the bearing journal average radius.

These measurements allow the computation of secondary data, such as the extent of the area of constant rising, the range of risings in the cam active zone and like. Since these computationsare trivial, they will be ignored in the sequel.

All measurements mentioned above are compared against theoretical data and a a quality certificate is issued. Moreover, coordination parameters for the machine tool used to process the camshaft can be produced. This information is of value in computer aided manufacturing, since it eliminates the human intervention in adjusting the improperly operating system component.

2. THE TESTING ALGORITHM

2.1. Working Hypotheses

The essential diagram in designing a cam is that of the tappet rising (TR) and, especially that of the tappet rising in the "active zone" (TRAZ). All other diagrams derive directly from this one. At this point first decision has to be taken in connection to the type of probe. We have chosen the plane probe. Of course, this might not allow the determination of the cam profile, but the tappet movement profile, and we considered that this is by far more important to know for the actual engine behaviour.

The TR is stipulated by the camshaft designers by a set of values (expressed in micrometres), corresponding to successive angular positions of the tappet in close contact to the cam. The term "successive" should be considered in connection to the camshaft rotation movement. Therefore, the introduction of the notion of "angular step" is appropriate , with the meaning of "angular interval between two successive rising values of the plane probe". Usually, the theoretical TR diagram (TRD) is expressed with an angular step of either 1^0 or 0.5^0. Consequently, the measurements will be effected with an angular step at most equal to that used in the theoretical TRD. A set of data composed of n values is associated to each of the N cams of the shaft:

$$x_i , 0 \leq i \leq n-1, \qquad (1)$$

in which n = 360 or an integer multiple of this value. The same notation can be used for the M bearing journals of the camshaft. Table 1 shows an example of a theoretical TR diagram.

Table 1 Theoretical TRD

Angle $\Phi[^0]$	Rising h$[\mu m]$
- 92	1
- 91	7
- 90	21
.....	......
- 2	4978
- 1	4993
0	5001
1	4993
2	4976
.....	......
89	9
90	1

For a full characterisation of the measurement space a system of cylindric coordinates (x, y, θ) is used (Fig.1), related to the measurement system and not to the camshaft itself.

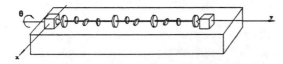

Fig.1. The cylindric coordinates system in connection to the measurement system

With the conventions established above, the tappet speed will be expressed in micrometres/degree, and its acceleration - in micrometres/degree^2.

The errors induced by the electronic measuring system will be neglected in the algorithm description in comparison with that of the mechanical part of the system, the accuracy of the latter being a problem lying mostly with the mechanics designer. Anyway, the performance of the computer based measurement system exceeds by far that achievable by any mechanical system. Therefore, the contribution of the electronics and software is negligible in the total error figure of the system.

Practical considerations prompt the choice of internal data representation. Indeed, since the values of x do not exceed a few tens of thousands of micrometres, 16 bit lenght will be used for input data.

2.2 Stages

It is obvious that data obtained from bearing journals are processed in a different manner as compared with data from cams. However, a testing algorithm (Ionescu, 1984), makes use of the computed bearing journal "excentricities" to produce, by linear interpolation, a set of correction values for the data obtained by cam measurement. This correction takes into account the real camshaft behaviour, which is rather different from that of the camshaft in the testing system, due to the presence of bearing caps in the real life situation. Due to this, the following stages are required in the process of measurement and testing:
a. measurement of bearing journals and parameter evaluation;
b. evaluation of cam excentricity with the view of obtaining correction parameters to be used in a later stage;
c. measurement of cams and evaluation of characteristic parameters.

It is also possible (Ionescu, 1987) to establish from the cam passive zone (if it is extended over more than 180^0) the cam excentricity without making use of bearing journal data.

One has to point out that he first two stages are optional, since they do not emphasise directly the cam parameters. By neglecting them, the test process takes much shorter time (the measurement of a bearing journal lasts as long as that of a cam, since the number of angular steps is the same and the angular speed is the same in both cases).

However, it is difficult to give a final answer to the question concerning the optimal data processing, namely what sets of data should be used to fully characterise the camshaft real life behaviour. This answer should came from the designers of mechanics, but we provided all facilities to take into account any possible set of measurements and parameters.

2.3 Processing of data associated to bearing journals

The computation of data obtained from bearing journal measurements is of little complexity. The starting hypothesis assumes that a bearing journal can be assimilated, ideally, to a circle of radius R and centre O, the latter being placed on the camshaft rotation axis. In reality, two types of errors can occur simultaneously: profile errors and, due to twisting and bending of the camshaft, excentricity errors (Fig.2).

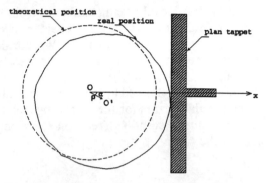

Fig.2. Theoretical and real life positions of a bearing journal

The "excentricity vector", expressed in polar coordinates (ρ, α), can be extracted from the set of measurements. The Fourier transform is used for this purpose (Stanasila, 1981). By using the notations (1) the following values are easily obtained:

(i) Average bearing journal radius:

$$\overline{R} = (1/n) \sum_{i=0}^{n-1} x_i \qquad (2)$$

(ii) The excentricity vector:

- module

$$\rho = \sqrt{u^2 + v^2} \qquad (3)$$

- angle

$$\alpha = \tan^{-1}(u/v) + k\pi \qquad (4)$$

$$k = \begin{cases} 0 \text{ if } u \geq 0, v \geq 0 \\ 1 \text{ if } v < 0 \\ 2 \text{ if } u < 0, v \geq 0 \end{cases} \qquad (5)$$

where:

$$u = (2/n) \sum_{i=0}^{n-1} x_i \cos(2\pi i/n) \qquad (6)$$

$$v = (2/n) \sum_{i=0}^{n-1} x_i \sin(2\pi i/n) \qquad (7)$$

2.4 Computation of cam excentricity

According to the remarks made in 2.2, the results of the previous stage (a) can be used to determine the cam excentricity, useful for correcting the data obtained directly through cam measurement. By using the linear interpolation, the excentricity vector for the cam C_l $(1 < l \leq n)$ can be computed. If $y_{l,j}$, $y_{l,j+1}$ are the cam longitudinal distances from the closest bearing journals P_j and P_{j+1} $(1 \leq j \leq M-1)$ the following formulae are derived:

$$\mu_l = \frac{\sqrt{\lambda_l^2 \rho_j^2 + \rho_{j+1}^2 + 2\lambda_l \rho_j \rho_{j+1} \cos(\alpha_{j+1} - \alpha_j)}}{\lambda_l + 1} \qquad (8)$$

$$\beta_l = \tan^{-1} \frac{\lambda_l \rho_j \sin \alpha_j + \rho_{j+1} \sin \alpha_{j+1}}{\lambda_l \rho_j \cos \alpha_j + \rho_{j+1} \cos \alpha_{j+1}} \qquad (9)$$

$$k = \begin{cases} 0 \text{ if } \mu_{l,x} \geq 0, \mu_{l,z} \geq 0 \\ 1 \text{ if } \mu_{l,x} < 0 \\ 2 \text{ if } \mu_{l,x} \geq 0, \mu_{l,z} < 0 \end{cases} \qquad (10)$$

where:

$$\lambda_l = \frac{y_{l,j+1}}{y_{l,j}} \qquad (11)$$

$$\mu_{l,x} = \frac{\lambda_l \rho_j \cos \alpha_j + \rho_{j+1} \cos \alpha_{j+1}}{\lambda_l + 1} \qquad (12)$$

$$\mu_{l,z} = \frac{\lambda_l \rho_j \sin \alpha_j + \rho_{j+1} \sin \alpha_{j+1}}{\lambda_l + 1} \qquad (13)$$

For the significance of notations refer to Fig.3.

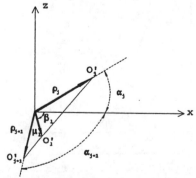

Fig.3. Significance of notations

2.5 Processing of data associated to cams

As Table 1 shows, the theoretical TRD contains data with reference to the 0^0 angle of the cam, corresponding to the so called "reference axis" and usually (although not always so) to the cam peak (maximum raising point). Therefore, in order to compare the real TRD against theoretical TRD it is first necessary to determine the cam reference axis.

The method we suggest to determine the angular position of cam reference axis uses measurement information obtained in the zero acceleration zones of the cam profile. The inlet and exhaust laws require the existence on the cam profile of such zones. They have various extensions, depending on the shaft type. Sometimes, these zones are symmetrical with respect to the cam peak. It is worth noting that the zero acceleration zones are also characterised by the maximum speed along the cam profile.

Given a set of data as in (1) associated to a particular cam and the characterisation of the zero acceleration zone by measure angle θ_i and the pair (Φ_i, h_i) and the raising x_i one can determine that the actual cam position is rotated (relative to the theoretical position) with the angle

$$\Delta\theta_i = \frac{z_i - h_i}{h_{i+1} - h_i} \; \theta_s \qquad (14)$$

Here θ_s is the angular step of the measurement. In order to simplify the treatment, θ_s is considered equal to Φ_s (the angular step between succesive theoretical values).

In Fig.4 the case $x_i > h_i$ is shown.

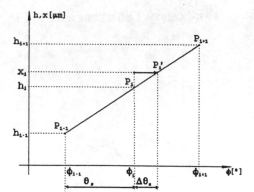

Fig.4. Interpolation of x in the zero acceleration zone

Since it is possible for x_i to rest outside one of the

expected intervals $[h_{i-1}, h_i)$ or $[h_i, h_{i+1})$ (i.e. $|\Delta\theta_i| > \theta_s$), a more accurate formulation for (14) is in order:

$$\Delta\theta_i = \left(\frac{z_i - h_i}{h_{i+1} - h_i} + q \right) \theta_s \qquad (15)$$

Usually, q is limited to values depending on the cam profile and imposed accuracy (as a rule: $|q| \leq 3$). The overtake of this limit must be cosidered as a serious error.

The real cam profile being different from the ideal profile, it is necessary to consider several points in the constant speed zone. We consider the same number m of points on both zones. The estimation of the cam angular position error (relation (18)) is derived by averaging the values of the angular errors for the ascending (relation (16)) and the descending (relation (17)) active zone of the cam:

$$\Delta\theta_A = (1/m) \sum_{k=1}^{m} \Delta\theta_{A,k} \qquad (16)$$

$$\Delta\theta_D = (1/m) \sum_{k=1}^{m} \Delta\theta_{D,k} \qquad (17)$$

$$\Delta\theta = \frac{\Delta\theta_A + \Delta\theta_D}{2} \qquad (18)$$

The number of chosen points depends on the extension of constant speed zones and, also, on the tolerance accepted for the cam rotation. The usual values for these numbers are 3, 5 and 7.

Once the value from (18) is obtained, interpolation is required in order to compared measured against theoretical data:

$$\overline{x}_i = x_i - (x_i - x_{i-1}) \frac{\Delta\theta}{\theta_s} \qquad (19)$$

If $|\Delta\theta|$ exceeds θ_s, a prior rotation of the data set is necessary.

After the rotation, (17) is applied to the new set of values. Following that, the values of $\overline{x}_i$ are compared to their theoretical correspondents, which permits to assess their position with respect to "tolerance band" (Fig.5).

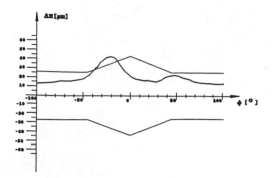

Fig.5. Behaviour within tolerance band

The essential cam parameters being determined, the other determinations follow suit:

-the angular offset between cams k and l:

$$\Delta\theta_{k,l} = \Delta\theta_k - \Delta\theta_l \qquad (20)$$

- the real speeds:

$$\overline{v}_i = \overline{x}_i - \overline{x}_{i-1} \qquad (21)$$
for i corresponding to "active zone"

- the real tappet accelerations:

$$\overline{a}_i = \overline{v}_i - \overline{v}_{i-1} = \overline{x}_i - 2\overline{x}_{i-1} + \overline{x}_{i-2} \qquad (22)$$

We emphasise that the suggested method takes into account the two essential elements in cam testing:

$$\Delta\theta \text{ and } \overline{x}_i, \qquad 0 \le i \le 359$$

4. THE TESTING SYSTEM

The testing system has been built around an IBM compatible PC, provided with an intelligent interface, the latter being able to provide the raw results of required measurements. This electronic system is connected to the mechanical system which allows the camshaft rotation and probe displacement. Obviously, the mechanical system has to be designed in accordance to the type of camshaft under test, while the electronic system remains the same in all circumstances. Note that the test system can operate also in manual mode (the movements are controlled by a human operator); this facility is required during commissioning operations.

The system easily allows its integration in a CIM scheme, due to the versatility offered by the computer. Under special circumstances, a dedicated equipment built along the general lines exposed above can be used.

5. CONCLUSION

The algorithm which constitutes the topic of the paper has been tested through implementing it on four systems dedicated to engine manufacturers in Romania. The results are of good value with respect to accuracy and efficiency of testing. The accuracy is better by an order of magnitude in comparison to that of the mechanical system. A full test takes less than 10 mins for an 8 cam shaft. The performance is limited by the use of low cost inductosyn type transducers for position measurement. With this kind of performance the connection of the automatic camsahft inspection and testing system into a complex CIM systemis straightforward.

6. REFERENCES

Stanasila, O., (1981). "Analiza matematica", Ed. Didactica si Pedagogica, Bucuresti.

Anon., (1982). "Samputensili SU 500/DAC User's Manual", Bologna.

Anon., (1982). "Studii referitoare la comportare a dinamica a camelor", raport tehnic, Institutul de Invatamint Tehnic, Pitesti.

Anon., (1983). "Studii referitoare la comportare a dinamica a camelor", raport tehnic, Institutul de Invatamint Tehnic, Pitesti.

Ionescu, T., Ceaparu, M. and Dascalu, S.,(1984). "Testarea automata a arborilor cu came de la autoturismele DACIA 1300", CNETAC 5, Bucuresti.

Carstoiu, D., Dascalu, S. and Moldovan, L., (1985). "An Application of Computer Aided Testing in Automotive Industry", Conference on Control Systems and Computers Science VI, Bucuresti.

Ionescu, T. and Dascalu, S., (1987). "A method of measuring camshaft parameters implemented on ESMC", Conference on Control Systems and Computers Science VII, Bucuresti.

Dascalu, S., (1989). "Aspecte constructive si functionale ale subsistemului de calcul, masura si comanda din componenta ESMCO4", Al II-lea simpozion de "Structuri, algoritmi si echipamente de conducere a proceselor industriale", Iasi.

MAKING CONFIGURATION EXPERT SYSTEM DEVELOPMENT LESS COSTLY

BERND G. WELZ

Institute for Real-Time Computer Systems and Robotics, Prof. Dr.-Ing. U. Rembold and Prof. Dr.-Ing. R. Dillmann,
University of Karlsruhe, Postfach 69 80, D-7500 Karlsruhe 1, F.R.Germany, E-mail: welz@ira.uka.de

Abstract. In this paper the software environment KONTEST is introduced which allows a non-programmer to build and maintain a knowledge-based configuration tool for a restricted class of configuration tasks. The use of an environment like KONTEST can be especially interesting for small and medium sized companies, that could not afford a customised configuration tool developed from scratch.

Key Words. Artificial Intelligence; expert system tools and techniques; knoweldge based configuration; knowledge acquisition.

1. INTRODUCTION

To meet the individual needs of the customers, a product, e.g. a machine tool, often has to be available in thousands of variants. To inform a customer about this variety and to help him find a configuration that meets his requirements best, is a difficult task that requires a great deal of technical knowledge of the sales person. Generally, this knowledge is not sufficient to validate the technical correctness of a configuration reliably, which makes a separate validation step of all orders by a technical expert necessary. With knowledge-based configuration tools the highly specific knowledge of few technical experts can be multiplied and made widely available throughout a company. For example, a large percentage of incoming orders can be validated from the sales person, thus saving processing time, insuring a high quality standard of the configurations sold and taking some of the load off the technical experts.

Unfortunately, as in other domains of expert systems, the knowledge acquisition, i.e. the process of elicitation, analysis and formalisation of knowledge, is a difficult and time consuming task. The maintenance costs cannot be neglected because due to short innovation cycles of the products the knowledge base of a configuration expert system has to be updated quite frequently. Unlike large companies, small and medium sized companies generally neither have the knowledge nor the resources to build and maintain such an expert system.

The purpose of the software environment presented in this paper is to enable non-artificial-intelligence-experts and even non-programmers (as most technical experts tend to be) to build and maintain a simple configuration system to a large extent without outside help. Building an expert system does no longer require costly external assistance or substantial additional training of staff. This could make way for the use of expert systems in an environment where it has been previously impossible for economical reasons, as for example in small and medium sized companies.

2. CONFIGURATION

2.1. Characterisation of configuration tasks

Configuration in a technical context denotes the process of putting together an object from a fixed set of elementary *components* that cannot be altered. The result of this process is also called *configuration*.

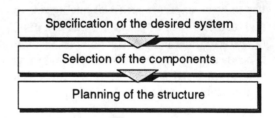

Fig. 1. Sub tasks of the configuration process.

In general, there are three different sub tasks in a configuration process (Fig. 1). In the first phase the desired system is described by a set of parameters or properties, then the components are selected according to these requirements. Typically, there are technical constraints, which limit the number of possible combinations of components. In the last phase a structure that can be built with the selected components is planned. Since in real world configuration tasks, not all sub tasks are equally important, configuration tools are therefore often restricted to one of the sub tasks.

2.2. Existing configuration tools

A large number of applied research in the area of configuration systems is reported in literature. The most prominent example of a configuration system is John McDermott´s R1, later called XCON (McDermott, 1982). XCON is certainly a milestone in the history of expert systems, but its purely rule-based approach to configuration is not suited to build easy to maintain configuration systems. Due to the lack of modularity of the rule-base and the homogeneous representation of knowledge with different roles, soon maintenance problems became apparent (Soloway, 1987). Only the impressive estimated benefits of XCON of more than 10 million Dollars per year justify the considerable effort to maintain the rule base of XCON: about 30 persons are employed for this task.

More recent work on configuration try to identify different roles of knowledge in the configuration process and find adequate representations for them. An example for this approach is the PLAKON system. PLAKON (Cunis et al., 1987) is a kernel for configuration and planning tasks in technical domains. A combination of two hierarchies, which denote *is-a* and *part-of* relations between components, and a constraint network describe the knowledge about valid configurations. The basic configuration mechanism selects components (or abstract components) by instantiating the corresponding elements in the hierarchies.

The separate representation of different kinds of knowledge à la PLAKON makes a configuration system much more transparent. Nevertheless, one has to be an expert in artificial intelligence to build and maintain such a system. Similar to a good piece of software, that is well structured and documented, it needs some knowledge about programming to be able to read and understand it and it needs a great deal of knowledge to be able to write it.

3. OBJECTIVES OF THIS WORK

The objective of this work, is to provide a software environment that helps build tools for a limited class of applications, but offers a maximum of support during the development and maintenance of the tools, to the extent that even non-programmers can use it. In particular, two requirements have to be met:
1. All formalisms, representations, etc. must be intuitively understandable.
2. The user must not be forced to formalise his knowledge, unless he already has a formal representation in his head.

Such a high level of support is only possible by designing a development environment specific to a restricted class of configuration tasks. Thus a great deal of knowledge generic to the task, the different kinds of knowledge needed and the control mechanism can be already included in this kernel. The knowledge that has to be added for a specific application is then clearly defined and can be elicited from the (application-) expert in a very focused way using dedicated tools.

4. FORMAL MODEL OF THE CONSIDERED CONFIGURATION TASKS

This section defines a model for the class of configuration tasks that can be supported by the system presented in the next sections. The model is best suited to describe the second configuration sub task, i.e. the selection task.

4.1. Definitions

A *configuration schema s* consists of a fixed set of *configuration variables* a_i:
$$s = (a_1, ..., a_n)$$

A variable a has a finite set of possible values v_i, the *domain* of the variable:
$$Dom(a) = \{v_1, ... , v_m\}.$$

A configuration c is complete when for all variables a_i a value $v_{a_i} \in Dom(a_i)$ has been specified:
$$c = (v_{a_1}, ..., v_{a_n}).$$

4.2. Interpretation of the model

This rather abstract terminology has been chosen, because the introduced concepts can be interpreted quite differently. A natural interpretation of our model is to see a configuration schema as an order form for a technical system. The order form consists of a number of slots (= variables). A slot on the form may correspond to an abstract functional unit, e.g. consider the order form for an assembly robot: one slot might be used to make a choice among different grippers. A slot may also be interpreted as a parameter. In our example, there might be a slot to specify the desired voltage of the power supply.

4.3. Conflicts

What makes configuration a difficult task is that not all configurations are correct, i.e. in our model, some variable-value combinations are not allowed. A configuration c is *consistent*, if it does not contain any invalid combinations of values.

A type of inconsistency is called a *conflict*. A conflict is an abstraction of individual instances of inconsistent situations to meaningful classes of inconsistencies. In the domain of robot configuration a conflict is for example "gripper not compatible with robot type", there might be a number of different gripper-robot combinations that are not compatible. Typically, not all variables in a schema are relevant for a conflict. In our example, only the variables "gripper" and "robot" are relevant. The set of relevant variables for a conflict k is called *conflict range R(k)*. Conflict ranges do not have to be disjunctive, i.e. a variable can be relevant for different conflicts.

5. DEVELOPMENT ENVIRONMENT

In this section the development environment KONTEST-D is introduced. KONTEST-D contains a number of functionalities that can help a user to acquire the knowledge needed for a configuration system based on the model presented in the previous section. The resulting knowledge base can then be used together with the run-time environment KONTEST-U (described in the next section) as an application specific configuration tool.

Two methods are offered to bring the knowledge about valid and invalid variable-value combinations into the system (Fig. 2).

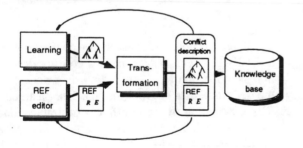

Fig. 2. Acquisition of conflict descriptions.

1. Using machine learning technology, the system can infer rules to detect inconsistencies from examples of correct and incorrect configurations. These rules can be automatically optimised and then further refined. The method is especially suited when the expert has no clear understanding under what conditions a certain conflict occurs. Thus he is not forced to give a formal description, if he does not already have one.

2. If the expert already has some idea about the reasons of an inconsistency he can directly express this knowledge in the so called *Rule-Exception-Format* (REF) which allows him to express even complicated relations in a natural way that can be intuitively understood.

The two methods do not have to be used exclusively. It is intended, that for example the expert first gives some examples to the system to get an initial description of a possible type of inconsistency, then he can further refine the resulting REF. Various refinements and optimisation functionalities are offered to allow the expert to build up, in an iterative process, a knowledge base that is correct and that makes sense to him. Especially, the last point is important to enhance the acceptance of the expert system and to facilitate maintenance.

As we will see in the next sections, a conflict has two representations in KONTEST: one suitable for an efficient evaluation of the conflict description and one for the user that should be intuitively understandable.

5.1. Learning of conflicts

To teach the system a new conflict, the expert first defines the conflict range of the conflict. Then he gives positive and negative examples for this conflict. A modified version of the incremental algorithm ID5R of Utgoff (1989) is used to learn from these examples. An incremental algorithm makes it possible to test the performance of the decision tree anytime during the training session. For a complete description of the algorithm and the ideas it is based on see (Utgoff, 1989) and (Quinlan, 1986).

The conflict descriptions generated by ID5R are represented as a decision tree (Moret, 1982). Formally, one can define a decision tree to be either: a leaf node that contains the decision "consistent" or "inconsistent", or a non-leaf node that contains a variable test with a branch to another decision tree for each possible value of the variable. A configuration is tested, whether a certain conflict occurs, by traversing the decision tree that is associated to this conflict.

The representation of the conflicts as decision trees allows to classify a configuration in linear time. Thus in the run-time system KONTEST-U the decision tree representations of the conflicts are used to check whether an inconsistency has occurred.

The extension of the original ID5R learning algorithm concerns the following. Typically, not all configuration variables in the conflict range are relevant for all instances, so the algorithm has been modified to handle *protypical examples*. In a protoypical example e of a conflict k not all variables in the conflict range are specified, i.e. some variable values are "don't cares":

$$e = (v_{a_1}, ..., v_{a_h}) \qquad \text{with } a_1, ..., a_h \in R(k)$$
$$\text{and } v_{a_i} \in Dom(a_i) \lor don't\ care$$

Since the number of possible values for every variable is known, the number of instances a protoypical example represents can be calculated. This number can then be used for the calculations performed by ID5R to decide if the structure of the tree has to be changed. In some cases, calculating the number of (elementary) examples is not sufficient, then the prototypical example has to be split up in a set of examples. (Namely, when the tree-update algorithm comes to a node that contains a variable test that is not specified in the prototypical example, it has to be split up among the children of the node.)

The possibility of not having to fill out examples completely saves time and allows to focus the attention to relevant relations between variable-value pairs. The number of examples needed depends on the structure of the conflict to be learned and on the quality of the examples given. Representative examples and examples, that show critical situations, will generally lead faster to a correct decision tree for a conflict than randomly selected examples.

5.2. Rule-Exception-Format

Typing in examples can be annoying for the user in cases, where he already has a clear understanding of a conflict. Especially, since the user has to give examples of both classes, i.e. consistent and inconsistent configurations. Apart from examples, the user should also have the possibility to specify directly the conditions under which a conflict occurs. For this purpose the *Rule-Exception-Format* (REF) has been developed. The semantics of the REF (Table 1) is modelled after the intuitive understanding of the concepts "rule" and "exception": a rule describes a fact, which is valid in most cases and invalid only in some exception cases.

The set of rules in REF describes all cases that are inconsistent concerning the conflict under consideration, but also some cases that are consistent. To define a conflict precisely, the user can specify exceptions that describe all cases that are misclassified by the rules. A configuration that has been classified "inconsistent" by a rule can thus be "made consistent" by an exception. In many cases the combination of these two sets allows a compact description of a conflict. The semantic of a decision tree is easy to understand for humans but the represented concept seems to be easier to comprehend in REF, despite its more complicated semantics.

Formally, a description of a conflict in REF consists of two sets: a set of rules R and a set of exceptions E. Rules and exceptions have the form of protoypical examples yet with different semantics. A rule r (or exception e) is called *applicable* ($\cong$) to a configuration c, if all in r (or e) specified variable-values are identical with the variable-values in c. This means that all configurations, to which no rule is applicable are indirectly defined as consistent concerning this conflict. But, if a rule is applicable to a configuration, it is not sure that a conflict is given, before the exceptions are evaluated:

c is inconsistent concerning conflict $k \Leftrightarrow$
$((r_1 \cong c) \vee \ldots \vee (r_p \cong c)) \wedge \neg((e_1 \cong c) \vee \ldots \vee (e_q \cong c))$
with $r_i \in R_k, e_i \in E_k$,

Figure 3 shows a simple example of a conflict range with three Boolean variables. A formula describes all inconsistent cases regarding this hypothetical conflict (true: the conflict is given) and the consistent cases (false: conflict is not given). This conflict could be expressed in REF with one rule and one exception as shown below.

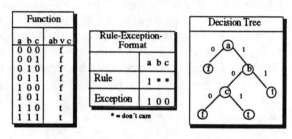

Fig. 3. Example of a Boolean function as a table, in REF and as a decision tree.

5.3. Transformations between REF and decision tree

To allow an efficient evaluation of a conflict defined in REF, it is necessary to have a conflict description as a decision tree. When the user wants to see the definition of a conflict that has been learned, a REF description of the conflict is desirable. Since the two knowledge acquisition methods introduced in the previous sections either generate one or the other representation, transformation algorithms are needed.

REF to decision tree: Since a conflict is completely specified in REF, there is no need for generalisation as in learning from examples, when transforming a REF into a decision tree. In (Welz, 1992) a simple algorithm is given that does not use any optimality criteria to generate minimal trees in terms of variable tests. The decision tree is reduced in a separate step. The tree reduction algorithm has been adopted from Cockett. In a recursive process, the algorithm tries to bring relevant variable tests near the root node. Variable tests, which are not in all cases significant for the identification of the conflict, are thereby pushed down towards the leafs of the tree or they are even eliminated, if they should be irrelevant for the subset of instances represented by the considered path of the decision tree. A detailed description of the algorithm, its limitations and complexity can be found in (Cockett, 1990).

Decision tree to REF: The transformation of a decision tree into REF is ambiguous. The goal is to find a description, which the user accepts because he can understand it easily. To find a description, which fulfils this requirement for all users, is not possible because of individual preferences of the users. The algorithm described here uses three heuristic criteria to measure the "goodness" of an REF description:

- K_R: number of rules -> should be minimised
- K_E: number of exceptions -> should be minimised
- K_*: number of "don't cares" in the rules -> should be maximised

The three partly conflicting criteria are normalised to values between 0, for the worst score, and 1, for the best score. The scores of a given REF are calculated on the basis of the structure of the tree. (Welz, 1992) describes the formulas for the calculation of the criteria measures.

To allow the user to express his own preferences, the final calculation of the overall goodness G of an REF d based on a tree t is parametrisable:

$$G = w_R \ K_R + w_E \ K_E + w_* \ K_* \ .$$

Similar to this measure of goodness, the search space for a good REF description is restricted using certain properties of the underlying decision tree. The algorithm only considers REF descriptions that fulfil the following requirements:

1. An exception e in d corresponds exactly to a path in the decision tree t that begins at the root and ends at a "consistent"-leaf. In e, the same variables are specified that have to be tested on this path.

2. A rule r in d corresponds exactly to a partial path in the decision tree t which begins at the root and ends at a "inconsistent"-leaf or at an arbitrary node.

3. The intersection of the sets of instances described by the rules (exceptions) is empty, i.e. there are no redundancies.

The quality of the REF depends thus heavily on the structure of the decision tree. The more compact the tree is, the better the REF can be. Despite these restrictions, the search still grows exponentially with the number of variables.

The algorithm to find an optimal REF description (according to the criteria described above) for a given decision tree t is recursive and starts with the root of t. First G for the REF d_r is computed. d_r is the REF, where the rule has been "placed" at the root of the tree. This means, d_r contains only one rule, namely, the rule which contains only "don't cares" as values, and a set of exceptions, that corresponds to the paths in t which end with a "consistent"-leaf. The result of this calculation g_r is now compared to g_s of the REF d_s, where d_s is the "best" REF, if the rule is not placed at the root. If g_r is greater than g_s then d_r is the better choice, otherwise d_s is the better REF description. The search for an optimal REF description is thus reduced for every subtree to the decision if a rule should be placed at its root or below.

If g_r is greater than g_s, then the root of t is marked as a rule-candidate. The algorithm traverses the entire tree once (in depth-first order), and this way marks all nodes as rule-candidates or non-rule-candidates. A node k in the tree t marked as rule-candidate means that the optimal REF description of the subtree with the root k contains only one rule, which is placed at k. In the optimal REF description of the entire tree t, a rule must only be placed at k if none of the predecessor nodes of k has been marked as a rule candidate.

To construct the optimal REF description of a tree according to the marks previously set, the tree is again traversed in depth first manner. When the algorithm comes to a node k which has been marked as rule-candidate the subtree is not further traversed and a new rule r_{new} is generated. r_{new} contains all variable-value pairs that are specified on the path from the root of t to k (including k). For every "inconsistent"-leaf, which exists in the subtree with root k, a new exception is created, which contains all variable-value pairs that are encountered on the path from the root of t to the leaf.

6. RUN-TIME ENVIRONMENT

The run-time environment KONTEST-U can load a knowledge base created by KONTEST-D. Since all conflicts are represented individually, also incomplete configurations can be tested, which allows KONTEST-U to give an on-line support of the user during the configuration process. Every time the user has specified a value, the system checks all conflicts (which conflict range contains the variable) as far as possible in the current state of the configuration process. If one of the conflicts is found, the configuration is inconsistent and the user is informed.

After the user has specified a number of values, he will typically find a desired variable value invalid in the current situation, because this value would cause a conflict together with the already specified ones. If he does not want to renounce this particular value, he has to change his choice for already specified variables. The system can support this process. It informs the user about which variables are relevant for this conflict, namely all variables in the conflict range. For a better understanding of the conflict, the system can also provide a verbal description of the conflict.

If the user is unable to find a consistent configuration or, after he has specified the values of variables, which he considers as important, he can have the system completing the configuration automatically. In the current implementation, KONTEST-U uses a backtracking algorithm for the search for a consistent configuration. The knowledge about the conflict ranges contributes to the decision which variables have to be changed to get a consistent configuration.

7. DISCUSSION, FUTURE WORK

The knowledge acquisition methods of KONTEST-D and the run-time environment KONTEST-U introduced in this paper have been successfully tested in artificial domains. In the near future, the system will also be tested in real world applications. Two possible domains are currently under consideration: configuration of computer systems and configuration of welding robots.

The results obtained so far have been promising. The combination of the inductive learning method and the direct implementation of knowledge by REF seems to be quite flexible. Also the two different representations of the conflicts seem to be appropriate. Some weaknesses in the transformation algorithms have been identified:

• Since the quality of the result of the reduction process depends on the original tree, more compact trees might be possible if the transformation algorithm produces a priori "better" trees.

• The algorithm decision tree to REF described in 5 is unable to find rules which reason about a disjunctive subset of the conflict range. Since the rules of a conflict normally involve approximately the same variables, this problem does not seem to cause major problems. Yet, with the size of the conflict range the need to represent such modular rules may increase.

The ongoing research is focused on refining the support of the formalisation process of knowledge following these ideas.

8. CONCLUSIONS

This paper introduced the system KONTEST, an environment for the development of knowledge based tools for configuration. It offers two mechanisms to acquire knowledge about possible inconsistencies of configuration, called conflicts: learning from examples and direct definition of conflicts in the Rule-Exception-Format. With the combination of the two mechanisms, a user is neither forced to formalise knowledge, which he has not readily available, nor he is forced to express his knowledge indirectly, where he already has a formalisation.

The REF seems to be promising as a formalism to represent complicated relations of variable-value pairs in a modular and for humans easy to comprehend fashion yet requiring only an intuitive understanding of its semantics.

9. ACKNOWLEDGEMENTS

This research work was performed at the Institute for Real-Time Computer Systems and Robotics, Prof. Dr.-Ing. U. Rembold and Prof. Dr.-Ing. R. Dillmann, Faculty for Computer Science, University of Karlsruhe, Germany.

10. REFERENCES

Cockett, J.R.B.; Herrera, J.A. (1990). Decision tree reduction. *Journal of the ACM*, Vol. 37, Nr. 4, pp 815-842.

Cunis, R.; Günther, A.; Syska, I. (1987). Planen mit PLAKON. *Tagungsbericht Workshop "Planen und Konfigurieren"*, Institut für Werkzeugmaschinen und Betriebstechnik, Universität Karlsruhe.

McDermott, J. (1982). R1: A Rule-Based Configurer for Computer Systems. *Artificial Intelligence* 13 (1982) 19, p. 39.

Moret, B. M. E. (1982). Decision trees and diagrams. *Computing Surveys*, 14, 593-623.

Quinlan, J. R. (1986). Induction of decision trees. *Machine Learning*, 1, 81-106.

Soloway, E.; Bachant, J.; Jensen, K. (1987). Assessing the Maintainability of XCON-in-RIME: Coping with the Problems of a Very Large Rule Base. *Proceedings of the AAAI-87*, Vol. 2.

Utgoff, P. E. (1989). Incremental Induction of decision Trees. *Machine Learning 4*, 161-186 Kluwer Academic Publishers, Boston.

Welz, B.G. (1992). Algorithmen zur Transformation RAS <--> Entscheidungsbaum, *Internal Working Paper*, WP-Welz-3.92/1.

Simulation in small and medium sized companies

G. Kronreif, P. Kopacek

Institute of Robotics, Technical University of Vienna, Moellwaldplatz 5/4, 1040 Vienna, Austria

Abstract: Modern production plants are characterized by enhanced complexity considering increased automation and enlarged integration of computers. Therefore duration and risks of planning are increasing. In addition putting into operation of new or modified plants bring about considerable complications with start-up operation and furthermore decreases in rate of production. The main source of this behavior is often the insufficient tuning of the single devices.

The aim of this paper is to give stimulus to utilize "simulation", which is well known as a planning tool for a considerable time. However it is being used too less in order to find solution to the problems mentioned above.

Keywords: Simulation; scheduling; modelling; manufacturing automation.

Introduction

Modern manufacturing and assembling plants are more and more characterized by the high degree of automatization of their single sub-systems. High integration of computers in these plants complicates production planning and increases the risk of mis-investment (Scharf, 1990). Moreover, the lot sizes are becoming smaller and smaller. Market's imponderableness however, asks for a higher flexibility in the production. By means of "Just In Time" JIT and "Computer Integrated Manufacturing" CIM some production or assembly cells cannot be decoupled by storage banks any longer, and one would not be able to correspond with the troughput times (Fürst, 1991).
Small and medium sized companies must therefore have high production flexibility to be able to meet the rapidly changing market situation, to meet the delivery dates and to provide for the increasing demands regarding quality.

Because of the reasons mentioned before quality and costs of planning contribute extremely the overall costs. At an early stage of production system planning, in particular, the costs of production are already largely determined depending upon the plant layout and the choice of machinery (Milberg, 1992). Costs can be saved considerably in this respect by means of an increase in the quality of planning. Because of the

decreasing life-time of certain products, another important factor, besides the reduction of planning costs, is the saving of time in the planning stage.

Conventional planning techniques cannot be utilized for this discussion. The results of modifications of the structure and the utilization of the system can only be estimated vaguely at this point. When looking for innovative and economic solutions instruments which show the weaknesses in an integrated process and which support the acquisition and realization of solutions and their utilization, have to be taken into consideration. Under this point of view simulation seems to be the most appropriate device. This technique makes it possible to prove the expected productivity of the planned, complex system. The system can be tested without interfering into the real process (VDI, 1983).

There are only a few applications of simulation in small and medium sized companies. One of the reason are the costs of simulation Technique.

First and second generation's simulation software and the partly wrong way of the development of a model led to the wrong opinion of an extremely costy and highly complicated technique. The new simulation systems with their graphical interfaces, their user-oriented modelling concepts and their time saving experimental environment are supposed to open new areas of utilization and to plead for more acceptance.

Advantages of simulation

Simulation is the imitation of the real plant in a computer model to deserve the dynamic behavior under several variants of load and eventual breakdowns. Based upon the costy and time-consuming experiments which result from the real process, simulation offers a lot of help in the development and modification of complex systems. The financial benefit of simulation-based planning must not be rejected already nowadays. The utilization of simulation technique permits a reduction of time spent in planning and of planning risk and an improvement in the quality of planning.

A positive side effect of the modelling is its extraordinarily high insight into the system. The plant is look at in a different way - from a different point of view. The formal description of the system, as it is necessary for the construction of a computer model, acts like a check-list (Fürst, 1991). In many cases, eventual errors are already discovered at this stage before the model itself is built up and existing but hidden reserves of the plant are recognized (Fig.1). Simulation in connection with animation of the manufacturing or the assembling process improve the communication between all the factors taking part in the process of planning.

Fig.1: Simulation generates a "Feed-Back" just during the phase of planning (Fürst, 1991)

New areas of utilization

A new sight of the simulation technique is growing increasingly. The simulation model is being serviced if possible improved and utilized from the phase of planning until the end of life-time of the manufacturing system. The renunciation of a one-way, so called "throw away" simulation and the acceptance of a long-term utilization of the simulation model built long ago leads to a saving of costs. It is worth, therefore, applying the simulation during the operating phase of the plant too, as a support of a PPS-System for example. In this case simulation contributes to a higher transparency and security in short-time planning decisions. It is possible to schedule an inserted order more detailed by means of the model of the planning phase and by the operating data derived from the actual situation. If an eventual change of products, a change within the collection or a modification of the sequence of operations might be required, this model determines these effects in advance and tries to find the adequate measures.

Simulation can therefore be utilized for several tasks (depending upon the time of employment):

- at the beginning stage of planning to discover realistic and dynamic relationships of capacities, costs, bottlenecks etc.

- during the planning phase to check the whole concept

- in the suppliance phase to check whether the means offered obtain the results requested before

- accompanying the phase of planning simulation may be used as testing environment to check the control software

- in the operating phase eventual effects of planned modifications can be determined in advance

- utilization of simulation in optimal scheduling

The continuing improvement of calculating speed and of the capability of computer graphics offers a further area of utilization.

The more and more realistic animation software makes it possible to use simulation as a means of support of marketing (WYSIWYG - What you see is what you get). Equipped with an extensive module sample, the seller profits by the fact that visual aid is better than lots of words. Because of the increasing appearance of powerful portable computers in particular simulation technique could be growing up more and more to a helpful tool during the marketing and selling phase.

Regardless of the utilization area of simulation, user-oriented software (graphical interface with interactive, menu- and window-driven modelling, block-oriented modelling concepts) are moreover preferred. Simulation can be utilized in those areas where it is requested - simulation is there for the planner and practitioner. Therefore, an initial period into the simulation technique and into the software used has to be offered to them which can easily be acquired by means of the new simulation software.

Costs and prejudices

Insufficient knowledge about this technique and its advantages

Beside the presumably high costs of the simulation technique its utilization is prevented by other factors as well.

The education program at schools and universities, for example, is far too limited. Moreover, people are still afraid of spending so much money on

Fig. 1.

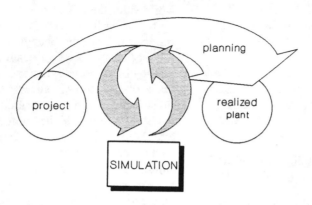

software or services in the software area. Finally, there is still this unfounded fear of this mighty and apparently complicated tool. "User-oriented" interfaces of the first and second generation's simulation software have been reinforced this fear. User-orientation an functionality combined with a vast variety of utilization are available only within the last few years in modern software packages.

Costs of simulation

From previous experience, the costs of simulation in general are up to 1-3 % of the total investment expenditure. It is difficult to understand in this respect why a technology with such a low expenditure and such a high economizing potential is not utilized more intensively.

In most cases the utilization of simulation in small and medium sized companies is prevented by high licence costs of the simulation software - efficient software packages require ATS 500.000,- on average. There are, however, other possibilities of how to get into simulation technique, like simulation studies as service, testing installation, run-time licence and much more. A step by step entrance is facilitated this way which leads to a reduction of the financial risk.

As was already mentioned, the turning away of a so called "one way" simulation, which means the utilization of the simulation model during the whole life time of the manufacturing system, can be regarded as a further contribution to the reduction of simulation costs.

Case study

As an application example serves an assembly cell for the production of primary parts of welding transformers in a typical medium sized Austrian company. Three years ago this company started with the realization of a "low cost" CIM concept including an automated assembling station. Because of the market conditions, the trend goes towards smaller lots and therefore more variations of the six basic types are necessary. Today approximately 1300 product variations with an average lot size of 10 are produced.

The necessary operations in the automatic assembling station are:

- attaching two transistors, four diodes, two resistors and one Thermocouple

- screwing these electronic parts

- attaching the printed circuit 1

- screwing the printed circuit 1 onto the diodes

- screwing three power cables to the printed circuit 1

- soldering the resistor cables and of some connections onto printed circuit 1

- attaching and soldering a capacitor onto printed circuit 1

- attaching a pressboard plate and the printed circuit 2

- screwing these two parts

The major part of the overall processing time is consumed by the technological tasks (i.e. time for soldering, screwing, etc.). Therefore the simulation could only handle the following domains (Fig.2):

- the sequence of the single assembling operations

- the single robot paths

- the optimal layout of the assembling station (i.e. the arrangement of the storage units, of the used grippers, of the screwing devices, etc.).

Fig.2:   layout of the assembling station and some of the robot paths (Kronreif, 1990)

The simulation, we used the simulation package SIMAN/CINEMA, has taken up approximately three months (Fig.3).

Fig.3:   snapshot of the simulation and animation (Kronreif, 1990)

The benefit of the simulation is the reduction of the processing time from once 6 minutes to 3 minutes now (for the smallest of the six different types).

Maybe this result could be reached without simulation too. Therefore they had to use the real process as an experimental enviroment. This way of optimizing - using the "trial and error" technique onto the real process - certainly is more time- and money consuming than using simulation technique.

Conclusion

Manufacturing industry is changing rapidly worldwide. Decisions have to be made more and more accurately and within shorter time of reflection. For this changing situation there is a tool at hand whose significant financial benefit has already been proved. Instead of using this tool for one's own economic success, one is afraid of its presumably high costs and runs the risk of investing a much higher amount in a wrong way. Simulation is an excellent instrument for experimenting promptly and cost saving. The simulation model and not the real factory serves as an experimental frame to optimize productivity and logistics. Simulation's utilization areas are transferred away from a single use of checking and controlling to a permanent function with the task of raising the benefit of the simulation model and saving time and costs for planning.

Simulation should be growing up to a key technology of the nineties. The trend to a user-oriented software, which can be used by everybody will be reinforced in the future. Simulation will be easier, more economical and more efficient this way. It must be a fixed component of the C-Techniques in planning in the future. Similar to the rapid development and distribution of these technologies, the CAD technology in particular, simulation will gain importance.

Fig. 2.

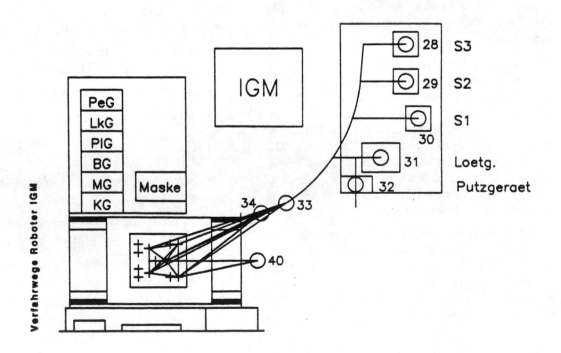

Fig. 3.

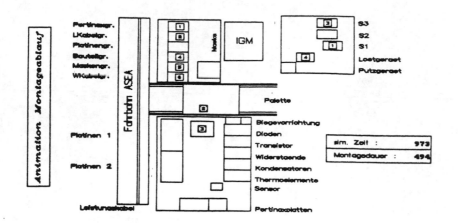

References

AESOP GmbH (1992). Simulation - a key technologie of the nineties: high benefit but insufficient knowledge in this technique. (in German) Fachaufsatz Simulation, Stuttgart.

Gangl P., and Hanisch R. (1992). A "guinea pig" made of bits and bytes. (in German) Lagertechnik, 11, 14.

Milberg J., and Burger C. (1991). Simulation as an aid for production planning und -control. (in German) ZwF 86, 2, 76-79.

Seevers C. (1991). Utilization of simulation for many tasks. (in German) Fördertechnik, 8, 62-65.

Fürst K. (1991). More sefety in planning because of Simulation. (in German) Fördertechnik, 7, 18-20.

VDI (1983). Simulations for material flow planning. (in German) VDI-Richtlinie 3633.

Milberg J., Amann W., and Raith P. (1992). Faster putting into operation because of tested sequence control. (in German) VDI-Z, 134/2, 32-37.

Scharf P., and Spies W. (1990). Simulation - results of an interview of several users. (in German) VDI-Z, 132/11, 62-65.

Kronreif G. (1990). Contribute to the simulation technique in the area of automated assembling (in German), Diploma Thesis, Institute of Robotics and Handling Devices, TU Vienna.

A LOW-COST PETRI-NETS BASED SIMULATOR
FOR SMALL SIZED MANUFACTURING SYSTEMS

A.M.STANESCU, S.I.BRADATAN, R.PATRAS, S.COSMESCU, G.COSMESCU

Polytechnic Institute of Bucharest, Computer Process Control Dept., CIM Division
313 Splaiul Independentei, 77206 BUCHAREST-ROMANIA

Abstract: The paper is dealing with a new approach for the integration of Petri Nets based simulator within a three level hierarchical software system to design the low-cost computer-aided planning of flexible manufacturing systems. This unconventional method is compared with the SIMAN language based mathod. The basic simulation procedure is presented as a two players events/actions Petri Nets modelling game.

Key-words: simulation, flexible manufacturing, Petri-Nets, modelling, scenario analysis

1.INTRODUCTION

The simulation plays an important role for both designing the new Flexible Manufacturing Systems (FMS) and Computer Aided Planning of manufacturing processes (Appleton, 1986). By modelling the FMS as a Discrete Events Dynamical System (DEDS) (Ho, 1990) - for example Petri-Nets based mathematical models - one could emulate this manufacturing process on the computer. The behaviour of the systems, in terms of time, should be observed without the necessity of experimenting with the actual equipments (Rembold, 1986).

Despite of the fact that the simulation is not mathematically equivalent to the optimization, however the iterative character of the simulator offers the facilities of searching for an optimum solution. So that, the main objectives of simulation are concerned with testing different manufacturing runs, balancing cell-oriented manufacturing systems, achieving the optimum layout of equipments, slowing down or speeding up cycle times, evaluating system performances, predicting the critical effects of external/internal disturbances. (Stanescu, 1992).

There are several different simulation methods and languages which are commercialy available. In principle, every high-level programing language, especially the object-oriented ones may be used to solve a simulation problem. However, the effort for developing a simulation package is costly affecting. Certainly, SIMPAS is used for the management of objects and states of least-driven

systems and, being written in Pascal, could be easily used for the definition of the initialization states and the halt of the simulation at any point. An event-driven time control could be succesfully implemented, as well as standard modules such as random number generator, the statistical distribution a.s.o. The simulation languages are currently clasified according to different concepts: event-oriented, process-oriented, transaction-oriented, activity-oriented (Rembold, 1986). By comparing GPSIS or SIMULA general use languages with SIMAN specific language is obviously the refinement trend, but the latter is rather expensive.

The main goal of this paper is to discuss a new approach of manufacturing system simulation based on three-levels hierarchical architecture. The Petri-Nets modelling is used to design the basic modules of simulation environement.

2. UNCONVENTIONAL VERSUS CONVENTIONAL APPROACH OF FMS SIMULATION

The concurrent engineering for CIM oriented enterprises is introducing a number of interesting problem which cannot be easily solved by conventional approaches, as the following issues:

- to design concurrently products and processes

- to speed up the cycle-time of production processes

- to use effective costly the software & hardware resources of the CIM system

• to share optimally the knowledge between human-network and Decision Support Systems (DSS)

The simulation, as a powerful software tool consisting of methods, activities and resources is deeply correlated with the global function of Computer Aidded Planning (CAP) within the CIM system.

As a starting point, the production flow analysis will be performed, the end being the simulation session.

The well-known conventional approach, described briefly in fig.1. uses oftenly SIMAN as the simulation language (Schulz,1990). Various modules are previously called for both introducing and analysing the production input data (workcell list, capacity requirements, processing times, batch size, transport, routing, a.s.o.). By processing these data, stored in a data-base, the optimum structuring of production flow will be provided by SIMILAR, CLUSTER, FERTPRINZ modules. The desired layout will be provided by COBAP module, whilst COSIM is interactively preparing the simulation program. Then, the syntax od the simulation model will be available in terms of relevant data.

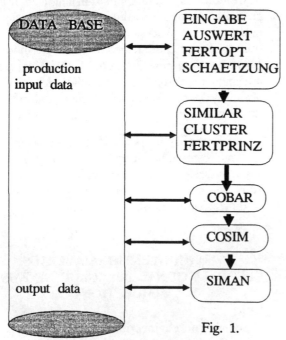

DATA BASE

production input data

EINGABE
AUSWERT
FERTOPT
SCHAETZUNG

SIMILAR
CLUSTER
FERTPRINZ

COBAR

COSIM

SIMAN

output data

Fig. 1.

The package is powerful and all the CAPP requirements are accomplished succesfully, but it is rather complex and expensive. The user training is requested for a special investment also.

Due to the complexity of the CAP problem, it seems to be natural to develop a hierarchical architecture for the system, each level receiving certain functions with respect to Increased Precision Decreased Intelligence principle (i.e. at the higher levels in the simulation scheme the accuracy is not as high as at the lowest levels, but the intelligence involved is increased).

In fact, the global goals of the control system are to be:

• **to optimize** the workcell configurations in terms of customary requirements

• **to provide** specific tasks for each "functional entity" and a corresponding time-schedule

• **to specify** the necessary syncronizing actions between functional entities in order to plan the actions, minimizing a cost-function

• **to perform** the low-level automation

Note: A functional entity is an autonomous equipment which is able to send, receive, process and (optional) store information.

The FMS configuration is changing because of both strategical criteria selection in terms customer's orders and fault rate for functional entities. So that, the configuration is to be oftenly re-built based on a set of procedural rules and on a knowledge base includig a diagnose expert system (ES). These facilities will predict and evaluate (as cost, duration a.s.o.) functional entities faults. It also seems to be naturally to provide a scenario of activities. There is possible to have several workcell configurations for a certain specifications set. To validate the couple (workcell configuration-scenario) the simulation is necessary.

Referring to these considerations, the proposed architectural structure of the control system include the following three levels:

Level 1: A high level diagnosis and reconfiguration level, establishing the optimim configuration and the best scenario for a given work-configuration.

Instead of a mathematical model, a behavioural description of the process is needed, based on qualitative expression and on the experience of the people involved in FMS supervising.

The choice of the best peer (work0configuration, scenario) is performed in two steps, supporting in a cooperative manner the level 2.

Level 2: A medium level application integration. Its goal is to simulate the scenario performed by the level 1, indicating the eventual faults and calculating the cost-function associate with the appropriate peer (work configuration,scenario).

Level 3: A low level flexible production automation, meaning contro program providing.

There are cases in which the evolutive scenario, based on Petri- Nets modelling has to be modified due to the occurence of unexpected sequences of events (external or internal disturbances). This problem needs to be detected during simulation sesion and a human expert should modify the scenario.

The system will carefully store the "history" of the process evolution to support the ooperator's decision.

Also, when a scenario fails at level 2, the operator could analyze the situation and make desired modifications.

The conventional approach is described in fig.2.

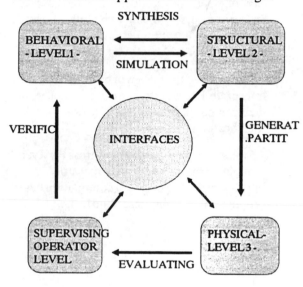

Fig. 2.

3. MEDIUM LEVEL PETRI-NETS BASED SIMULATOR

By foocusing the attention on the simulation subject, one could formulate the following design problem:

Input Data :

Scenarios from level 1, consisting of actions flow specifications for workcells and various subsystems (transport, handling, warehousing, buffering, a.s.o.) involved in the manufacturing process.

The model taked into account for the manufacturing process is behavioral, so that the scenarios are provided in terms of events/actions, as well as the time duration for every subsystem

Goals to be performed:

- to verify if the actions flow is feasible (no blocking state for example)

- to estimate the minimum/maximum duration of a certain scenarioo

- to evaluate the cost function

- to prepare, from a successful scenario, specifications for every cell-controller, as partitioning the application programs for every functional entity

- to introduce within the scenario the "emergency routines" for the case of functional entity failures

Output Data:

- Feasibility reports delivered for level 1

- Program specifications for level 3

Due to the behavioral modelling of the FMS/FMC Petri Nets are appropriate as simulating formal entities (David, 1989). Aiming our goals, Interpreted Petri Nets (IPN) were chosen, due to the following reasons:

- IPN are synchronized

- IPN are place-time delayed

- the basic entity includes an operational term, having $V = \{V1, V2,...\}$ as a state vector. The state may be modified by the operational vector $O = \{O1, O2, ...\}$ which are associated at a certain place.

- The actual state is determining the value of predicate set $C = \{C1, C2, ...\}$ which are associated at a certain transition.

In fig.3. the symbolic peer place/transition is shown. The delay di as well as the operational vector Oi are associated at the place Pi, whilst the Ej event and Cj predicate are associated at the Tj transition.

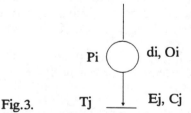

Fig.3.

The Tj transition will be activated **if** transition Tj is valid **and** predicate Cj is true **when** (and only when) event Ej occurs

In order to simplify the synthesis phase, a two-step simulation is proposed (Stanescu, 1991)

First step: Simulation of level 1 macro-scenario The activities inside a cell are not considered. Every cell is modelled as a macro-entity. Its environmental interactions are enhanced by this macro-modell. If this rough simulation is successful, then the second step will be performed

Second step: By using a powerful module-library, the operator will perform a detailed Petri-Nets modell for every FMC (macros). The interactions among cell components (fuctional entities) are introduced.

4. BASIC SIMULATION PROCESS DESIGN

The basic simulation process is designed as a two-players game. First player is the Petri-Net modelling the scenario of activities (either for FMS macr-simulation or for each FMC-detailed simulation). The second player is the behavioral model (also Petri Nets based) of the manufacturing considered process.

Informations are changen between the two players via a common register, storing all the variables considered to be events/actions for the two players. These variables are set up at the begining of the simulation by initial conditions values.

A time clock is also set-up at the zero-moment. An auxiliary memory is storing the time varying variables to allow a hystorical analysis if it is necessary.

Player 1 reads the register by searching for significant events (state vector change). If there are occured, player 1 performs the required transitions, modifies register variables corresponding to actions and then allows player 2 to take the program under control.

Player 2 also searches for variables which are conditions or which generates events for it and produces an adequate set of inputs, modidying the associated varables stored in the commom register.

The game is going on, until player 1 finds again its initial state vector and/or join the desired end of the scenario.

If player 1 fails, then a failure report is generated for level 1, optionally specifying the conditions and history of the failure.

Else the second step of the simulation will be performed by a sublevel (2.2) of level 2 based on the specifications of the initial scenario. The sublevel 2.2 will calculate the cost-function for every macro-module. The results are send back to sublevel 2.1 to calculate the global cost-function, the time-interval for the scenario development. These results will be reported to level 1.

5. CONCLUSIONS

To promote the lean production for small sized enterprises, some basic investtigations for cost effective use of software resources of CIM.E system have to be undertaken.

It is the main goal of the present paper to present a new approach of the FMS simulation problem. A higher modularized three-level hierarchical environment is proposed for CAP requirements. Scenario analysis is going to be followed by the iterative two-steps simulation in order to minimize a cost-function.

As concerning the design of basic modules, the Interpreted Petri Nets modelling is chosen due to their compatibility with events/actions specifications.

An intensive work is under progress to develop the routines library as well as the operator interface. Further goal is to perform a compilation based on proposed philosophy described in fig.2

6. REFERENCES

Appleton, D. (1986). Introducing the New CIM Enetrprise Wheel" - CASA/SME Publications Development Dept., Deaborne

David, R., Alla, H. (1989) Du Grafcet aux Reseaux de Petri - Ed. Hermes, Paris

Ho, C.Y. (1990). Analyzing complexity and performance in a manmade world - An introduction to discrete events dynamic systems - The First IFAC World Congress, Tallin, Estonia

Schulz, H. (1990). - Integriertes System zur Fabrikplanning - PTW Dokementz, Darmstadt

Stanescu, A.M., Bradatan, S.I. (1991). Computer Aided Planning of robot-driven welding cell - Technical Report granted by Science Dept. / Ministry of Education, Bucharest (in Romanian)

Stanescu, A.M. (1992). Computer Integrated Manufacturing Systems Topics - Polytechnic Institute of Bucharest Press (to be published in Romanian)

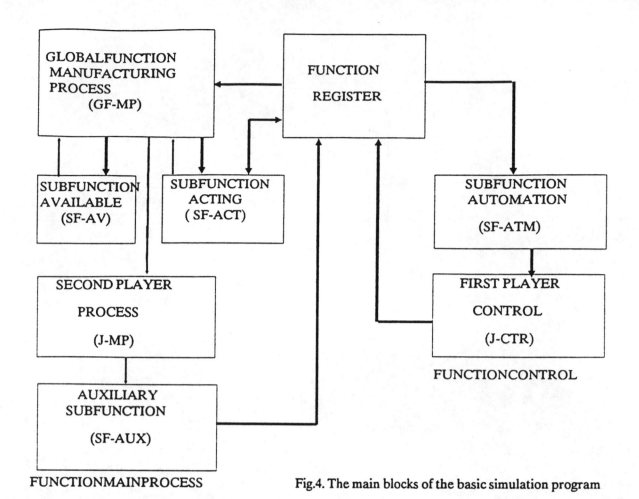

Fig.4. The main blocks of the basic simulation program

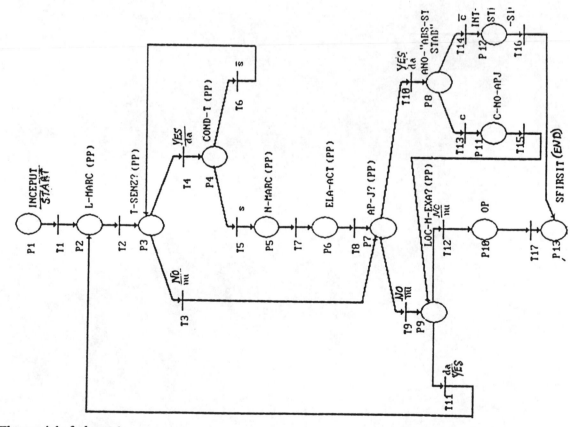

Fig.5. The model of player 1

153

A LOW COST MEASURING CELL FOR CIM ENVIRONMENT

P. Herbert OSANNA, N. M. DURAKBASA, R. OBERLÄNDER

Vienna University of Technology - T.U.Wien, Austria
Karlspl. 13/3113, A-1040 Wien

Abstract. The application of CAD, CAE and CAM also puts new demands on inspection and testing in computer integrated production processes. A sophisticated measurement technique, among other things, is considered a most crucial requirement for the production of industrial goods of a controlled and optimized quality. The past twenty years have seen a continued increase in importance of computer aided measurement techniques as a means to control industrial production. At the same time the basic definitions governing design, production and quality control of workpieces have undergone considerable advance with the goal of international harmonization and standardization. Focal points of interest included the definitions of nominal, real and substitute geometrical features. Plane, sphere and cylinder are examples where the method's divergence depending of the point of view - this can be design, production or measurement - still causes trouble in industry.

Key Words. Production metrology, geometrical features, computer integrated production, quality management

1. INTRODUCTION

Competition and cost consciousness on the one side and increasing demand for quality and reliability on the other side are contrary requirements in modern production engineering. This must be seen also under the point of view of the new international standards about quality assurance and quality management [1].

In the last few years, the standards governing product design and manufacture have undergone basic international harmonization; focal points of interest included surface roughness [2] and deviations in form and position [3], as well as tolerancing principles according to the principle of independence [4]. In many countries, the above mentioned international standards have been adopted also on a national level.

It is often necessary to prescribe tolerances of size, form, position and roughness if specific workpiece accuracy is demanded. The knowledge of necessary and permissible deviations from the ideal geometry of workpieces excercises great influence on the economics of manu-

facture. Because of lack of knowledge the design engineer or the man at the CAD system pescribes very narrow tolerances which are the reason of unnecessary expensive production costs.

As already established by Kienzle [5] there is evidence for the existence of "inherent interrelations", as it were, among the different geometrical deviations of workpieces.

Depending on whether macrogeometry or microgeometry is in the focus of the analysis of workpiece surfaces, a distinction is made between form errors of different order, both in the technical literature and the relevant standards. It is common practice to collectively consider the more or less short wave geometric deviations of a third or higher order as surface roughness on the basis of worldwide understood and internationally established parameters [2].

As far as geometrical deviations of form and positions are concerned the increased use of co-ordinate metrology has improved common knowledge [6 to 9].

There exists a series of influences that must be taken into consideration. It is general knowledge to take into account the ISO tolerances [10 and 11] and general tolerances [12] respectively for linear dimensions. But also geometrical deviations of form and position must lie within certain limits whereas general tolerances [13] or tables of experimental values respectively [14] can help the design engineer who has to specify allowances on drawings.

Workpiece accuracy is also affected by surface roughness and therefore it is necessary to specify corresponding parameters in drawings. On the one hand there is an influence on measurement conditions, on the other hand it has been shown that accurate co-ordinate measuring machines can be used to evaluate workpiece microgeometry [15] instead of special devices.

2. MODERN PRODUCTION METROLOGY

If it is necessary to choose a measuring device it is also necessary to know the reasons of possible deviations in so far dimensional as well as form measurements should be done. Only with three co-ordinate measuring machines (CMM) it is possible to measure deviations of dimensions, form and position very accurate with only one measuring device. Besides measuring accuracy the number of workpieces to be measured is important when choosing the measuring device. Especially when workpiece tolerances are more accurate than tolerance grade IT5 it is necessary to make use of co-ordinate metrology. This is also possible for big series of workpieces.

In general we can define co-ordinate metrology as follows:

- The geometrical features of the workpiece to be measured are touched in various measuring points using a co-ordinate measuring device.

- The co-ordinates of the measuring points are used to compute the mathematical geometry of the workpiece with help of the computer of the co-ordinate measuring machine.

At the time being co-ordinate metrology is a very important tool to solve various problems of pro-duction metrology especially when high flexibility and high accuracy are demanded.

When we consider as example a bore hole we see the difference between conventional measurements and the application of co-ordinate metrology. A two-point-measurement has to be done by means of the strategy trying to get the maximum value perpenticular to the axis and then finding the minimum in the axial section (Fig. 1). The bore diameter is given by the condition that these both diameter values are equal.

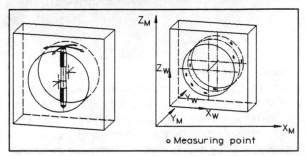

Fig. 1: Conventional metrology and coordinate metrology

In contradiction to this method we get the diameter of a bore hole when using co-ordinate metrology by contacting the surface in several measuring points whereas at least five points have to be measured because of mathematical conditions. When we measure more than five points we must use regression methods for the evaluation of the measuring result. But in addition to the dimensional result we also get informations about the deviations from the ideal cylindrical form and about the position of the feature respectively in a workpiece co-ordinate system dependent of the overall shape of the workpiece (Fig. 2).

In this context it is distinguished between nominal, real and substitute geometry [6, 7, 8]. This follows the way of workpieces from design to manufacturing and quality control (see Fig. 2). When we look on workpiece microgeometry [3] we can distinguish in a very analogous way (Table 1).

Table 1:

Co-ordinate Metrology	Workpiece Microgeometry
nominal feature	geometrical surface
real feature	real surface
substitute feature	reference surface

156

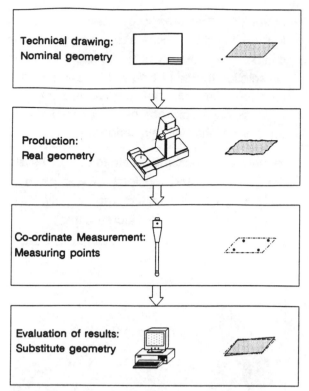

Fig. 2: Nominal, real and substitute geometry

measurement problems in an environment of computer integrated production. This should be achieved with a minimum of manual operations. Such a flexible measuring cell is an important part of a modern system of quality assurance (Fig. 3).

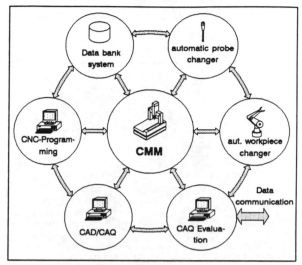

Fig. 3: Flexible measuring cell, main tasks, components

So these ideas can be seen as a further step in view of the integral approach according to Weingraber [17] and the interrelations between different deviations from ideal geometry shown by Kienzle [5].

If we take into consideration cylindrical features another distinction can be made. On the one hand it is distinguished between Gauss and Chebyshev regression element as far as best fitted elements are concerned, on the other hand we are interested in the minimum circumscribed and the maximum inscribed element if the feature must fit to a counterpart.

3. MEASURING SYSTEMS IN CIM

CAD/CAQ datacommunication for quality management in computer integrated production is possible on the basis of co-ordinate metrology.

A measuring cell for the flexible automation of measurement, quality management, data collection and data evaluation in small companies and industrial plants was drawn up at the Vienna University of Technology [18]. When we try to give a definition: A flexible measuring cell is a compound of hardware and software with the task to find flexible solutions for all kinds of

4. FLEXIBLE MEASURING CELL

There are various tasks that must be solved by this measuring system:

- automatic measurement by using CNC measuring programms,

- offline CNC programming of the CMM,

- automatic changing of workpieces,

- automatic probe changing,

- automated evaluation of measuring results.

Fig. 4 gives a block diagramm of the described flexible measuring cell. The system consists of the following devices and components:

- a precision CNC co-ordinate measuring machine with control computer and printer for measuring results,

- a probe changer with interface and control computer,

- a robot for workpiece manipulation,

- a local area network of various personal computers especially for CAD, CAE and CAQ evaluation,

- various printers and plotters for data and graphic output,

- various measuring instruments, for instance roughness and form testers, a small CMM, a laser scanner and other devices,

- data bank systems for construction data, technological data, measuring results and quality data.

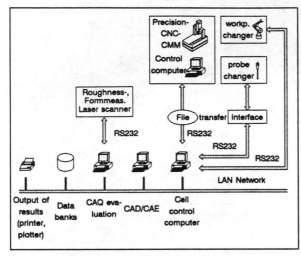

Fig. 4: Flexible measuring cell, block diagramm

The application of this measuring cell and the proposed solution can be seen as a further step with the goal to achieve economical manufacturing, inspection and management of quality data in modern industry especially in small and medium sized plants. With great flexibility this results in an accurate and wastefree production. This is shown in detail by the example of cylindrical workpiece features and new solutions for the measurement related geometrical definitons can be given.

5. CAD RECONSTRUCTION

To disengage the control computer of a CMM for new measuring tasks measured data in the form of co-ordinate values of the measuring points of workpiece features are transferred to an evaluation computer. Especially when working with many data for instance in the case of exact form measurements this is essential for economic work. For the CMM only collects measuring data and the extensive and longer lasting evaluation is done outside of the measuring devices.

In addition measured values can be transferred to a CAD system whereas graphical judgements can be done very obviously. To find an economical solution this can be solved by using the serial interface and a personal computer for evaluation. Special software must be used for data transfer and further calculations.

By transfer of measuring data to the CAD system the dimension and the form of workpiece features can be interpretted very clearly. Fig. 5 shows as example a measured cam.

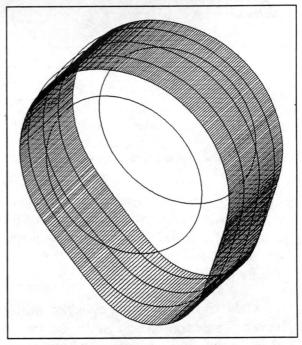

Fig. 5: CAD reconstruction of a measured cam

6. CONCLUSIONS

The data transfer to an evaluation computer is of special importance when extensive calculations of different regression elements are necessary.

The proposed concept of a flexible mesuring cell can be seen as a further step with the goal to achieve accurate and automated inspection and quality management in computer integrated production on the basis of low budget systems. Besides linking of the CMM and the various computer systems it is also essential to provide for communication with data banks for geometrical allowances. With great flexibility this results in an accurate and wastefree production.

7. REFERENCES

[1] **ISO 9004-1987:** Quality Management and Quality System Elements - Guidelines.

[2] **ISO 1101:** Technical Drawings; Geometrical Tolerancing; Tolerances of Form, Orientation, Location and Run-out. Generalities, Definitions, Symbols, Indications on Drawings. 1983

[3] **ISO 4287-1:** Surface Roughness - Terminology - Part 1; Surface and its Parameters. 1984

[4] **ISO 8015:** Technical Drawings; Fundamental Tolerancing Principle. 1985

[5] **Kienzle, O.:** Formtoleranzen. Werkstattstechnik und Maschinenbau 45 (1955), No.11, pp. 605/607, p.615

[6] **ISO/TC3/WG10 N39:** Dimensional and Geometrical Co-ordinate Measurements. Part I: Terms,Definitions. Geometrical Fundamental Principles. Code of Practice. May 1988

[7] **ON M 1380:** Koordinatenmeßtechnik. Geometrische Grundlagen. Grundlegende Benennungen und Definitionen. Austrian Standard, 1988

[8] **DIN 32880, Teil 1:** Koordinatenmeßtechnik. Geometrische Grundlagen und Begriffe. Draft Standard, Dec. 1986

[9] **ANSI/ASME B89.1.12M-1985:** Methods for Performance Evaluation of Coordinate Measuring Machines. 1985

[10] **ISO 286-1:** ISO System of Limits and Fits - Part 1; Bases of Tolerances, Deviations and Fits. 1988

[11] **ISO 286-2:** ISO System of Limits and Fits - Part 2: Tables of Standard Tolerance Grades and Limit Deviations for Holes and Shafts. 1988

[12] **ISO 2768-1:** General Tolerances; Tolerances for Linear and Angular Dimensions Without Individual Tolerance Indications. 1989

[13] **ISO 2768-2:** General Tolerances; Geometrical Tolerances for Features Without Individual Tolerance Indications. 1989

[14] **Osanna, P.H.:** Deviations and Tolerances of Position in Production Engineering. Wear Vol. 109 (1986), No. 1/4, pp.157/170

[15] **Osanna, P.H., Durakbasa, N.M.:** Comprehensive Analysis of Workpiece Geometry Using Co-ordinate Measurement Technique. Surface Topography 1 (1988), pp.135/141

[17] **Weingraber, H.:** Einheitliche Definition, räumliche Messung und eindeutige gegenseitige Abgrenzung von Formabweichung, Welligkeit und Rauheit. Feingerätetechnik 25 (1976), No.2, pp. 58/62

[18] **Durakbasa, N.M., Osanna, P.H., Cakmakci, M., Oberländer, R., Waczek, G.:** Flexible Meßzelle. e & i (Elektrotechnik und Informationstechnik) 108 (1991), No.6, pp.257/261

AUTHORS:

Univ.Dozent Dr.techn. P.Herbert OSANNA is head of the Department for Interchangeable Manufacturing and Industrial Metrology (Austauschbau und Meßtechnik) at the Institute of Production Engineering, Vienna University of Technology (T.U. Wien), Austria, Dr.techn. Dipl.-Ing. Numan M. DURAKBASA and Dr.techn. Dipl.-Ing. Rudolf OBERLÄNDER are members of the staff.

A Modular "Low Cost" CAQ System

M. Zauner*, J. Hölzl** and P. Kopacek* ***

Dept. of Systems Engineering and Automation, Scientific Academy of Lower Austria, Krems

**Fronius Company, Quality Assurance, Wels, Austria*

***Institute of Robotics, University of Technology, Vienna, Austria*

Abstract: Quality demands under the pressure of international competition can be satisfied only specific for small and medium sized enterprizes with respect to the standards ISO 9000 to ISO 9004. The descriped and self developed CAQ system consists of 20 modules. All of them are integrated in an existing CIM concept. The CAQ system has been considered at creation time of the existing CIM system. Therefore a relational CIM data base in addition to the PPS database are connected to the CAQ data base. The modular software is arranged in layers and supports the client/server principle. A network (ethernet) with MS-DOS PC´s and Unix-Workstations are the main hardware components.

Keywords: CAQ, CIM, modular CIM concept, client/server architecture.

1. INTRODUCTION

Since 30 years quality has become more significant in international competition. The high quality level of products manufactured in central europe is caused in the well develpoed education system, specially in the skill based production. Nevertheless increase in quality is required for maintaining competition. The accomplishment of quality demands must be realized according to the standards of ISO 9000 to 9004, whereas the very specific structures of the companies have to be taken into account.

The importance of quality increases compared to to terms like costs, efficience and date. It is necessary to introduce measures for quality assurance. In order to obtain orders quality-certifications will be nessecary prerequisite for enterprises. For Austrian enterprises the new european economical market underlines these requests.

2. THE IMPORTANCE OF A CAQ-SYSTEM FOR SMALL AND MEDIUM SIZED COMPANIES

The use of CAQ-systems is very significant especially for small and medium sized companies. For small and medium sized companies, which are producing in the supplying industry the use of quality assuring measures is already condition for relations to the mass producing industry.

When quality assurance systems will be introduced the requests of both, the market and customer as well have to be well defined. Depending on the inner structures of a company, all measures have to be taken in order to meet and assure the demands. An important component for planning, control and ensurance quality is to develop an computer aided quality system (CAQ).

A CAQ-system collects quality data about long periods. So deviations from tolerance in the production can be recognised and future quality tests can be planned, depending on errors, which occured in past. Beside of these operative advantages CAQ-systems offer implicit and difficult measurable advantages; for example, reduction of production errors can be reached by specifying standards and improving job planning.

It is very difficult to assign these benefits to particular cost reductions.

Considering the fact, that 75 percent of all errors are rooted in the development phase, then the potential for saving costs by use of CAQ-Systems is enormous.

The costs of errors mulitply from development via production to the commercial use of the product with the factor of 10. Therefor the avoidance of errors which are recognised in the development phase costs 1 money unit, in the production phase already 10 and in the commercial use 100 units.

At the same time, when failure costs are reduced, product quality, image, efficiency and meeting deadlines are being increased.

3. QUALITY PLANNING

Since 1975 CIM strategies have been developed. In this period substantial efforts were made to manufacture quality.

Today the goal is first to plan and afterwards to produce quality. Methods like Taguchi help to find failures in an efficient way (orthogonal arrays) before producing the product, methods of Shanin help to analyse failures after they have occurred. The methods of Taguchi and Shanin are often named a philosophy, because the classical way to run test routines for analyzing data is based on statistical methods like SPC. In the opposite to SPC (statistical process control), which is placed in production and, for example, inputs data to the methode of shanin, the FMEA is placed in the development.

In the domains of mechanical and electronical manufacturing the planning of quality by means of methods of taguchi and shanin and FMEA (Failure Mode and Effects Analysis) can be supported very well. The high expense to make FMEA´s in the development phase is justified by recognising many bottlenecks and trouble spots in an early phase of the product life cycle. The main goal of the FMEA is to analyse products and product parts before manufacturing and to ensure quality preventive. Important product parts are nessecarily recorded with FMEA in quality assurance systems.

FMEA is also an interface to the customer, because the detailed specification of the parts is carried out with the FMEA. Internal know-how should be kept out of FMEA before sending to the partner.

Hence, the goal of planning quality, to minimize deviations from the desired value, can be reached by using planning methods in addition to more interrelated software modules and high degree of network performance. Software modules support data acquisition, data administration and cost and method analysis. Considering the fact, that maschine data, production data, order and personal data are available in the company, computer aided quality assurance systems are important tools for planning quality and controlling the domains from design to final tests.

These research directions to improve quality in the actual working step as well as in the following working steps exists already in easy-to-assembly design (robotics) and in design for testability especially in the domain of chip and print development.

4. AN INTEGRATED, MODULAR CAQ-SYSTEM IN AN EXISTING CIM-SYSTEM

Aims

CAQ-systems are highly integrated modules within existing computer integrated manufacturing systems. This is due to special organisation structures and the degree of compatibility of the used software packages.

The domain of the software interfaces is low developed as well. The potential to "build" standards is enormous.

These problems exist in both, the large and the medium sized companies. Although these enterprises and software development research centers work hardly to solve these problems, the success increases slowly. There are hardware components and operation systems, which are too heterogeneously. Therefore it is a diffucult task to build interfaces between software packages. Apart from these factors, the commercial influence is also an important factor to assess this development.

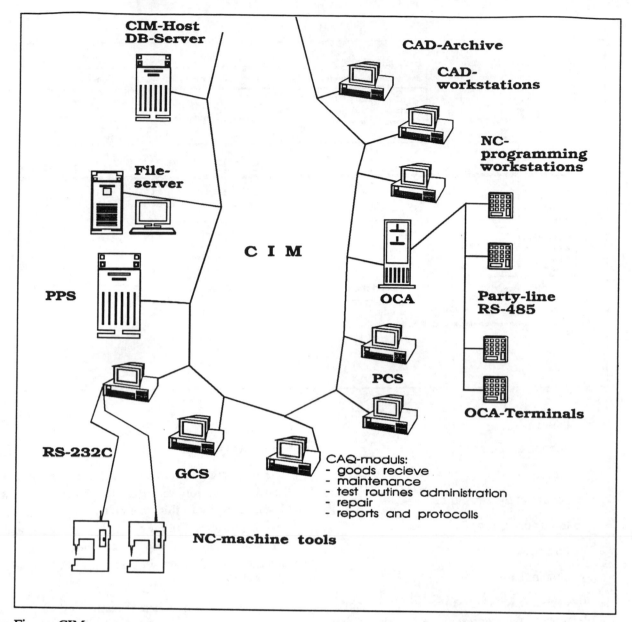

Figure: CIM components

In addition to these facts particular organisation structures and interfaces have to be taken into account by developing an integrated modular CAQ-system. Such a CAQ-system is a very specific product and today not available as a standard product.

In our special topic, a computer aided quality assurance system was planned, developed and implemented in an Austrian medium sized company in addition to the domain of the quality system.

This CAQ-system is integrated in an already existing CIM-system. This CIM system was developed for the demands of small and medium sized companies.

The company has about 650 employees.

The company developes and manufactures electronic and electro-mechanic parts and assembles them to welding transformers in diverse variants.

More than 17.000 single parts have to be administrated. Among these parts a high percentage is purchased.

The products are manufactured in small and medium lot sizes. So the number of orders within a period of a week is up to 600.

70 percent of all produced welding transformers are exported. So, the international competition and the economic stagnation has increased noticeably. Therefore, when setting up a CIM/CAQ-system, it is the objective to maximize

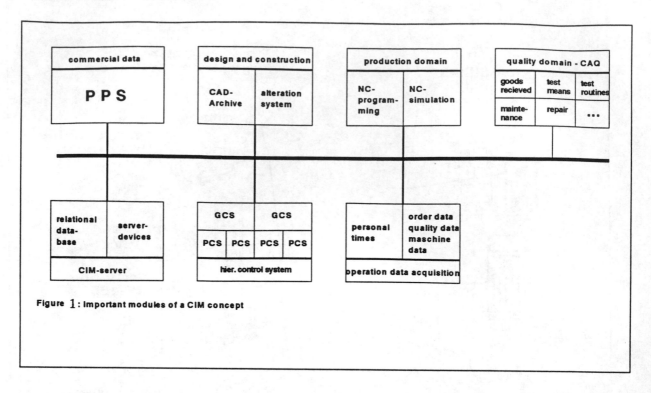

Figure 1: Important modules of a CIM concept

quality,

meeting dead-lines

efficiency

and to reduce at the same time

manufacturing costs

failure costs,

complaint costs.

In this company a CIM-concept was already implemented and represents a computer aided connection between planning and operative level. Classical CIM components as CAD and PPS have been complemented with modules like a CAD-archive, an alteration system, NC-programming tool and many more /Figure 1/. The modules hierachical process control system, operational characteristics acquisition (OCA), a module to administrate modifications, CAD/NC-archive, NC-programming, tool administration and CAQ are the most important self developed compontents of this CIM concept. Because the components PPS and CAD have already been existing before realizing the CIM concept they were used and integrated, but they have not been self developed.

The CAQ system

The use of a central, relational data base was a basic assumption in the framework of the development of this computer aided quality assuring system. The modules of this CAQ-system are completely integrated into the CIM-system and therefore they are provided with the interfaces to the CIM components.

With the central data base a common data pool is available with a uniform interface and standardized data access functions. The complexity during the implementation and the maintainance expense of such a system, especially in the administration of the interfaces is being reduced by the use of a central data base and standardized data access functions.

With the introduction of this CAQ system occurrence of failures should be reduced preventively. A module SPC was not developed because the lot sized in the production of this company are too small. The way to controll quality in the production is to analyse failure cost in the working groups.

The CAQ system includes 20 modules according to the 20 elements of quality assurance in ISO 9001.

The most important modules of the CAQ system are:

maintenance
goods recieved, order
repair
complaints
test routines administration (production, goods recieved, test means)
dynamical generation of test orders
FMEA
interim and final tests
collection and administration of test data
reports and protocolls

Table.: modules of the caq system

The quality data and data streams project specific states and informations of the company. To increase the reuseability of these domains, a modular conecpt was applied when the CAQ system was designed.

The caq-modules are set up uniformly. The screen masks are devided into three areas:

1. module information and state description

2. data domains

3. function menues

The software architecture is set up level by level, modular and supports the clients server concept. The data access functions to the Informix data base are self implemented and are called data communication management (DCM). They follow the standard of the ISO/OSI protocoll. With this concept the stepwise introduction of an integrated CIM/CAQ system can be realized.

Hardware components

In order to guarantee an optimal use of the hardware components the database and network server is a UNIX-workstation. All clients are implemented on PC's 286-386. The network is based on ethernet. Units like OCA-terminals and test means are connected to the PC's via serial RS-232 and RS-485. Test data can be input into the computer data base both via the display in some modules (goods received) or via OCA-terminals (production).

Experiences

Classical methods from software engineering as well as simultaneous engineering methods have been applicated for the development of these applications.

In the preparation period all modules and the most important facts have been defined. They have been sorted by priorities, whereas 4 levels of priorities have been used and each of them was tested according to the economic efficiency.

The analysis phase was devided into two steps. First of all it was necessary to collect and record all information and data flows of the company. After the interfaces have been defined prototypes of the modules were developed.

With the use of these prototypes the detailed analysis has been accomplished.

After the prototypes of the modules with highest priority level had been accepted, the specifications and programming activities started. The experiences, which we gained from these modules, could be transferred and used in the development of the next modules.

To develop the prototypes of the CAQ modules, the graphical interface tool and the simulator were utilized. As a result the screen masks could be developed and tested.

Software tools

For the development, implementation and testing of the CAQ-system following software tools were used:

relational database management system INFORMIX

graphical interface and simulation module GRIT

data communication management routines (DCM)

programming language C

The introduction and observance of company specific **standards** in the domains of development tools, data administration, program design, programming, and documentation reduced the expense for implementation to 50% in time.

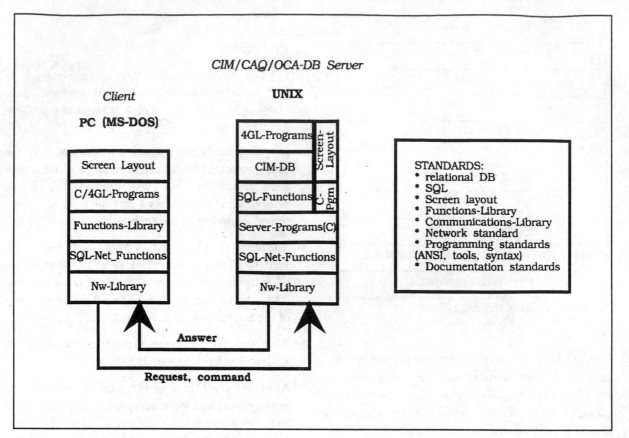

CIM/CAQ/OCA-DB Server

- Fig. 2.: Software architecture

Essential advantages were short feedback both to the customer and to the programmer team. Since the customer was involved in the development of the prototyp, he knew already his program.

For the programmer it was an inportant advantage to have an interface in the programming language C available to the graphical interface tool.

5. Literature:

Kopacek P., Frotschnig A., Zauner M.: "CIM for small companie"s, Automatic Control for Quality and Productivity - ACQP ´92, IFAC-Workshop, Proceedings, Istanbul 1992.

Kopacek P., Girsule N., Hölzl J.: "A low cost modular CIM concept for small companies", Information Control Problems in Manufacturing Technology - INCOM ´92, IFAC-Symposium, Proceedings, Toronto 1992.

Zauner M., Kopacek P.: "Data integration in CIM concepts for small and medium sized companies", Computer Aided Technologies - CAT ´92, User Congress, Proceedings, Stuttgart 1992.

553 2662

James Blyth Combe

E 8, 1/2

Bangor